T0302014

EQUATIONS OF PHASE-LOCKED LOOPS
Dynamics on Circle, Torus and Cylinder

WORLD SCIENTIFIC SERIES ON NONLINEAR SCIENCE

Editor: Leon O. Chua
University of California, Berkeley

WORLD SCIENTIFIC SERIES ON
ONLINEAR SCIENCE

Series A   Vol. 59

Series Editor: Leon O. Chua

EQUATIONS OF PHASE-LOCKED LOOPS

Dynamics on Circle, Torus and Cylinder

Jacek Kudrewicz

Warsaw University of Technology, Poland

Stefan Wąsowicz

Technical University of Czestochowa, Poland

World Scientific

NEW JERSEY · LONDON · SINGAPORE · BEIJING · SHANGHAI · HONG KONG · TAIPEI · CHENNAI

Published by

World Scientific Publishing Co. Pte. Ltd.

5 Toh Tuck Link, Singapore 596224

USA office: 27 Warren Street, Suite 401-402, Hackensack, NJ 07601

UK office: 57 Shelton Street, Covent Garden, London WC2H 9HE

British Library Cataloguing-in-Publication Data
A catalogue record for this book is available from the British Library.

EQUATIONS OF PHASE-LOCKED LOOPS
Dynamics on Circle, Torus and Cylinder

ISBN-13 978-981-277-090-5
ISBN-10 981-277-090-9

Printed in Singapore.

Preface

The Phase-Locked Loops are electronic systems which have numerous applications, such as synchronized oscillators, dividers or multipliers of frequency, modulators or demodulators and amplifiers of phase modulated signals.

This book is devoted to the study of nonlinear dynamics described by classical equations of these systems. Different mathematical models are discussed: continuous-time and discrete-time systems of the first and second order. Differential or recurrence equations of phase loops depend on parameters, mainly on the frequency and amplitude of the input signal. Particular attention is paid to investigate how these parameters influence the occurrence of various types of stable periodic output signals. Much space is devoted to the chaotic oscillations appearing in the system. It is not our purpose to study various schemes and applications of phase loops because these problems have already a rich literature. This book draws the attention of the reader to nonlinear physical phenomena which cannot be explained by approximate theory which uses linearized or averaged equations. Phase loops have rich dynamics, probably more exciting than Chua's circuit or Lorenz equations. Selected mathematical methods (theory of one-dimensional mapping of the circle, integral manifolds, bifurcation theory and other) are presented and applied to explain the qualitative properties of nonlinear oscillations.

This book is addressed to postgraduate students, researchers in nonlinear science and mathematically inclined engineers interested in dynamical phenomena, particularly in the phenomena of deterministic chaos.

The authors wish to express their thanks to their colleagues from the Institute of Electronic Systems at Warsaw University of Technology for creating a friendly atmosphere which made it easier to write this book. Special thanks are to Mr. Zdzisław Michalski for his help with computer experiments. We would also like to thank Prof. Maciej Ogorzałek for inspiration to undertake this work.

Jacek Kudrewicz   and   Stefan Wąsowicz

Contents

Chapter 1

Introduction

1.1 What is Phase-Locked Loop?

The phase-locked loop (PLL) is an electronic system which has numerous important applications.

It consists of three elements forming a feedback loop: voltage controlled oscillator (VCO), phase detector (PD) and low-pass filter.

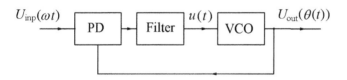

Fig. 1.1 Scheme of the phase-locked loop.

Generator VCO produces electrical oscillations $U_{\text{out}}\big(\theta(t)\big)$ periodic with respect to θ. The waveforms of these oscillations may be different, for example sinusoidal, saw-tooth, rectangular or other. The angular frequency $\frac{d\theta}{dt}$ depends on slowly varying signal $u(t)$ controlling the oscillator. Usually, if the controlling voltage increases, the generator frequency also increases.

The aim of the phase detector PD is to produce a signal which controls the generator. This signal depends on the phase difference $\theta(t) - \omega t$ of the generator signal and the input signal U_{inp}. Usually PD is a system which multiplies both these signals. If, for example, $U_{\text{inp}} = U \sin\big(\omega t + \varphi_1(t)\big)$ and $U_{\text{out}} = \cos\big(\omega t + \varphi_2(t)\big)$ then at the output of the phase detector we get

$$U_{\text{inp}} \cdot U_{\text{out}} = -\frac{1}{2} U \sin\big(\varphi_2(t) - \varphi_1(t)\big) + \frac{1}{2} U \sin\big(2\omega t + \varphi_1(t) + \varphi_2(t)\big). \quad (1.1)$$

If $\frac{d\varphi_1}{dt} \ll \omega$ and $\frac{d\varphi_2}{dt} \ll \omega$, the first term of the right-hand side of (1.1) is

a slowly varying signal used to tune the VCO, whereas the second high-
-frequency component is harmful and should be suppressed by a low-pass
filter.

The feedback system tends to minimize the error signal $\frac{1}{2}U \sin(\varphi_2 - \varphi_1)$
and, in consequence, the output signal phase tracks the changes of the input
signal phase.

The basic property of the PLL, useful in practice, is its ability to syn-
chronize (to lock in). It means that when the input signal is periodic then,
in a steady-state, the mean frequency of output signal generated by the
VCO is equal to the input signal frequency, and the phase difference of
these signals is small and almost constant.

There are two basic kinds of phase detectors. The first of them compares
the phases continuously, and the second does it only in discrete moments
of time determined by the output (or input) signal phase. The phase detector is then a sampling-and-holding system and the controlling voltage
$u(t)$ changes (jumps) at the moments when the output (input) signal phase
equals an integer multiple of 2π. The magnitude of this change depends on
the input (resp. output) signal phase. Such systems are used, among other
purposes, to multiply and divide the frequency.

In technical literature on the PLL systems (see Refs. [9], [10], [19], [22],
[28], [33], [34], [35], [53], [58], [59]) one can find a long list of various applications. The most important of them are:

1. Locked oscillators and frequency synthesizers,
2. Modulators, demodulators and converters (AM, FM, PM),
3. Recovery of the clock signal,
4. The tracking filters,
5. Frequency multipliers and dividers, coherent transponders,
6. Synchronization of digital transmission.

The PLL systems are produced also in integrated version and find use
in many technical devices of electronics and telecommunication.

1.2   PLL and differential or recurrence equations

The phase-locked loop with continuous time is most frequently described by
the system of ordinary differential equations. The order of these equations is
greater by one then the order of the filter which connects the phase detector
and the voltage controlled oscillator. In the state vector of the system the
first coordinate is the phase $\theta(t)$ of the output signal $U_{\text{out}}(\theta(t))$, the second

is the voltage $u(t)$ controlling the VCO, and the remaining coordinates characterize the inner structure of the filter.

In particular, if the filter is of the first order then the equations of the PLL system are of the second order

$$\frac{d\theta}{d\tau} = F_1(\theta, u, \tau), \qquad \frac{du}{d\tau} = F_2(\theta, u, \tau), \qquad (1.2)$$

where $\tau = \omega t$.

For the phase-locked loop without filter the equation is of the first order

$$\frac{d\theta}{d\tau} = F_0(\theta, \tau). \qquad (1.3)$$

The input signal $U_{\text{inp}}(\omega t)$ is most frequently a periodic function of time of the period $\frac{2\pi}{\omega}$. Then the functions F_0, F_1, F_2 are 2π-periodic with respect to the variables θ and τ. This specific property of equations enables us to treat the variables θ and τ as cyclic variables determined up to a multiple of 2π. To visualize the solutions of equations we then choose a space with suitable geometric properties (suitable system of coordinates). For the equation (1.3) it will be a surface of torus instead of plane, and for equations (1.2) torus or cylinder (solid).

Equations of each physical system depend on a number of parameters. In PLL systems two parameters are especially important: the amplitude and the frequency of a periodic input signal. So, it is necessary to investigate how the whole family of differential equation solutions corresponding to all admissible initial conditions depends on these parameters. Further investigations concern the dependence of solutions on constructional parameters of the system: the cut-off frequency of the filter, the coefficient of amplification and the quiescent frequency of VCO. It is not a simple task because these equations are strongly nonlinear and their solving demands using the advanced theory of differential equations.

The number of essentially different physical phenomena observed in the second order system (1.2) is greater then that occurring in the first order system (1.3). However, in some ranges of parameters the high order system behaves "just the same" as the second order system (and similarly, the system of the second order may behave like a system of the first order). This happens when there exists a globally attracting stable integral manifold of a suitably low dimension. The conditions for the existence of such a manifold have substantial practical significance and this is why we devote them a lot of place (Secs. 3.5, 3.9, 5.3).

In technical literature one usually admits many simplifications. The most important of them is neglecting the fast-varying part of the input signal of VCO. For the phase difference $\varphi(\omega t) = \theta(t) - \omega t$ (but not for $\theta(t)$) we get the autonomous system of equations

$$\frac{d\varphi}{d\tau} = F_{a1}(\varphi, u; \omega), \qquad \frac{du}{d\tau} = F_{a2}(\varphi, u; \omega), \qquad \text{where} \quad \tau = \omega t, \qquad (1.4)$$

instead of equations (1.2). The frequency ω occurs here as a parameter. This procedure of averaging will be justified in the next section. Of course, the equations (1.4) are easier to analyze then the equations (1.2), (compare Secs. 2.2, 3.2).

The second simplification is a linearization of the system in a neighborhood of the stable steady-state (compare Sec. 3.9.4). For the autonomous system the linearized equations are most frequently solved using the Laplace transformation and the so called transfer function. Transfer function has simple physical interpretation and provides electrical engineers with valuable information about the behavior of the linearized system. However, the linearization of the system permits to investigate only its local properties in a small neighborhood of the steady-state, but it gives no answer to such an important question as the range of initial conditions and parameters for which the system remains in synchronization (pull-in and hold-in ranges).

For discrete-time phase-locked loops the output signal phase $\theta(t)$ is compared with input signal phase $\tau = \omega t$ only in discrete moments t_n, for example, when $\theta(t_n) = 2\pi n$, for $n = 0, 1, 2, \ldots$ The behavior of the second order system is completely determined by the properties of the point sequence

$$(\tau_0, u_0), \quad (\tau_1, u_1), \quad (\tau_2, u_2), \quad (\tau_3, u_3), \ldots \qquad (1.5)$$

where $\tau_n = \omega t_n$ and $u_n = u(t_n)$. Instead of the differential equation the system is described by the recurrence equation (map), which relates a position of the point (τ_{n+1}, u_{n+1}) with position of the point (τ_n, u_n) in a plane or on a cylindrical surface if we treat the variable τ as cyclic variable determined up to a multiplicity of 2π.

For more simplified models of phase-locked loops instead of (1.5) it is sufficient to investigate the sequence of points

$$\tau_0, \quad \tau_1, \quad \tau_2, \quad \tau_3, \ldots \qquad (1.6)$$

on a straight line or on a circle (with perimeter normalized to 2π).

From the behavior of sequence (1.5) or (1.6) one can deduce whether the PLL system synchronizes, when there is multiplication or division of the input signal frequency, what the sensitivity of the system is with respect to small perturbations, how fast a synchronization state is reached, and also, when chaotic oscillations appear.

1.3 Averaging method

In technical literature the high-frequency component of the output signal of PD (the second term on right-hand side of (1.1)) is usually neglected. The motivation of this fact is purely physical: the high frequencies are damped by the low-pass filter and, therefore, have inessential influence on the behavior of the system.

There is also a formal mathematical reasoning (called the averaging method) for such simplification of equations. Below we present the averaging method and its range of applications.

Let us consider the differential equation

$$\frac{dz}{dt} = F(z, \omega t), \quad \text{where} \quad z = \{z_1, z_2, \ldots, z_n\}, \tag{1.7}$$

with positive parameter ω which takes a large value. Let

$$F_a(c) = \lim_{T \to \infty} \frac{1}{T} \int_0^T F(c, \tau) d\tau \tag{1.8}$$

denote the mean value of the function $F(c, \tau)$ with respect to τ.

The autonomous differential equation

$$\frac{dz_a}{dt} = F_a(z_a) \tag{1.9}$$

is called the *averaged equation* with respect to the original (1.7).

The solutions of the averaged equation (1.9) approximate solutions of the original equation in the following sense.

Theorem 1.1. *Let the following assumptions be satisfied:*

1) $F(z, \tau)$ is uniformly continuous and bounded function for all z and $\tau > 0$,

2) the limit (1.8) is uniform with respect to c,

3) the equation (1.9) with an initial value $z_a(0) = z^*$ has exactly one solution $z_a(t)$ for $t \in [0, \infty)$.

Under these assumptions, for arbitrary numbers δ (small) and t^* (large), there exists ω^* such that for all $\omega \in (\omega^*, \infty)$ the solution $z(t; \omega)$ of equation (1.7) with the initial value z^* satisfies the inequality

$$\max_{0 < t < t^*} ||z(t; \omega) - z_a(t)|| < \delta, \tag{1.10}$$

where $||z||$ denotes the norm of vector z.

The proof is omitted. Theorem 1.1 is one of many specialized theorems about solutions of a differential equation depending on a parameter. One can find a rich collection of similar theorems in Ref. [42]. A substantial part of the proof (see also Ref. [31]) uses Ascoli's theorem on compactness of any set of uniformly bounded functions equicontinuous on a bounded closed domain.

From the physical point of view the parameter ω is a frequency of periodic signal perturbing the physical system. According to the above theorem, if the frequency is sufficiently large then in any finite time interval the system behaves almost the same as the autonomous system in which the time-varying perturbation is replaced by its mean value. However, replacement of the finite time interval $[0, t^*]$ by infinite interval $[0, \infty)$ is generally not possible (if δ decreases or if t^* increases then ω^* increases). A small distance between the solutions of both equations, i.e. (1.7) and (1.9) in an infinite time interval (in a steady-state) occurs rather exceptionally and under much stronger assumptions.

One of the theorems on the averaging method related to steady-state signals is formulated below.

Theorem 1.2. *Let the assumptions* 1) *and* 2) *of the theorem 1.1 be satisfied, and moreover:*

3) $F(z, \tau)$ *is a periodic function of* τ,

4) *there exists* z_0 *such that* $F_a(z_0) = 0$,

5) *the Jacobian matrix* F_a' *is continuous in a neighborhood of the point* z_0 *and it is non-singular at the point* z_0 *(i.e.* $\det F_a'(z_0) \neq 0$).

Under these assumptions there exists ω^* *such that for all* $\omega \in (\omega^*, \infty)$ *the equation* (1.7) *has in a neighborhood of* z_0 *exactly one periodic solution* $z_{\mathrm{per}}(\omega t)$ *(of the same period as* $F(z, \omega t)$) *and*

$$\sup_{-\infty < t < +\infty} ||z_{\mathrm{per}}(\omega t) - z_0|| \to 0 \quad as \quad \omega \to \infty. \tag{1.11}$$

Moreover, if all eigenvalues of the Jacobian matrix F_a' *have negative real parts at the point* z_0 *then the periodic solution* $z_{\mathrm{per}}(\omega t)$ *attracts neighboring solutions (it is asymptotically stable).*

The proof is omitted (the implicit function theorem is used in the proof).

It is worth a mention that in mathematical literature the theorems on averaging are usually formulated for equation of the form:

$$\frac{dz}{d\tau} = \varepsilon F(z, \tau), \quad \text{where } \varepsilon \text{ is a small parameter.} \tag{1.12}$$

This equation becomes identical with equation (1.7) after time rescaling $\tau = \omega t$ and changing the parameter $\varepsilon = \frac{1}{\omega}$.

Averaged equations of phase-locked loops are solved in Secs. 2.2, 3.2. However, conclusions which follow from these solutions have limited applications. The behavior of solutions of averaged equations in a long time interval (of the order of several hundreds of input signal periods) may be qualitatively different from the behavior of solutions of original PLL equations, where the frequency ω is a fixed parameter (and is not a value which one can freely increase). An extensive part of this book deals with comparing solutions of original equations with solutions of averaged equations. In particular we investigate how the high-frequency component at the input of VCO changes such solutions of averaged equations as, for example, equilibrium points, stable periodic trajectories and separatrices of saddle points (the borders of attractive domains).

1.4 Organization of the book

The purpose of this book is an investigation of nonlinear deterministic models of PLL systems when the high-frequency term of the signal at the input of VCO is not neglected, but to the contrary, its essential influence on the system's behavior is emphasized. Subsequent chapters concern continuous-time systems of the first and second orders as well as discrete-time systems also of the first and second orders. In this way we pass from easier problems to more difficult ones accepting a few repetitions of some small parts of the material. Each chapter can be read almost independently. The sections which present concrete conclusions concerning the behavior of PLL systems are preceded by sections which include the necessary mathematical notions and theorems with the proofs. Some proofs are omitted (when they can be found in most textbooks, or when they concern less important theorems).

Chapter 2 presents a nonlinear model of phase-locked loop without filter, described by differential equation on a torus. The solutions of averaged Adler's equation are investigated. The notions of rotation number, devil's staircase and Arnold's tongues are introduced as basic characteristics of synchronized oscillators which depend on parameters. This material shows the mechanizm of multiplication and dividing the frequency (fractional synchronization) in the most simple phase-locked loop systems. The cases of sinusoidal and rectangular waveform signals are discussed in the first place.

In Chapter 3 we are dealing with the second order phase-locked loop system with a low-pass filter. The phase-plane portrait of an averaged system on a cylinder is examined, and conclusions concerning hold-in range and pull-in range are presented. The considered system is periodically perturbed by high-frequency component of input signal. We investigate what happens with some selected trajectories (fixed points, stable periodic orbits and separatrices which connect saddle points) after the perturbation. The conditions for the existence of a stable one-dimensional integral manifold are given. Under these conditions the dynamics of the second order system is reduced to dynamics of the first order system. The Melnikov theorem is applied to examine different types of homoclinic trajectories, and their influence on the transient chaos and on the form of attractive domain borders of stable fixed points. The last Sec. 3.9 is devoted to systems with higher order filters. A theorem on the existence of a globally stable two-dimensional integral manifold for the equations of a phase-locked loop of higher order is formulated and proved. The conditions are given for which a higher order system is reducible to the second order.

In Chapter 4 a discrete-time phase-locked loop is investigated. A continuous two-modal mapping of a circle is accepted as the mathematical model of this system. The properties of periodic points, rotation intervals, frequency locking regions and attractive domains of stable periodic orbits of different types are discussed. The emphasis is on bifurcations of periodic orbits (saddle-node and period doubling bifurcations), Feigenbaum's cascade, the skeleton of superstable orbits and bifurcation of rotation interval on the border of a frequency locking region. Some characteristics of chaotic dynamics, especially the invariant measures and Liapunov's exponent are also discussed.

Chapter 5 deals with a more realistic model of discrete-time phase-locked loop, described by a two-dimensional continuous map of the cylinder. Stable periodic points and hold-in regions in the plane of parameters are examined. The conditions for the existence and decay of a one-dimensional stable invariant manifold are investigated. Different types of attractors are discussed, e.g.: stable periodic orbits, invariant curves and strange attractors. In particular, the relations between homoclinic trajectories and bifurcations of strange attractors (crisis bifurcation) are given. Transitions from stable chaotic oscillations to a transient chaos are illustrated by numerical experiments.

Chapter 2

The first order continuous-time Phase-Locked Loops

2.1 Equations of the system

The first order phase-locked loop (PLL) contains two elements: a voltage controlled oscillator (VCO) and phase detector(PD).

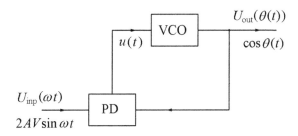

Fig. 2.1   The first order phase-locked loop.

Let U_{inp} and U_{out} be two 2π-periodic functions. Examples of these functions are shown in Fig.2.2. The output signal of VCO is $U_{\text{out}}\big(\theta(t)\big)$, where instantaneous frequency $\frac{d\theta}{dt}$ depends on the controlling voltage $u(t)$ in the following way:

$$\frac{d\theta}{dt} = \Omega\left(1 + \frac{u(t)}{V}\right). \tag{2.1}$$

This equation has two parameters: Ω – the quiescent frequency of VCO and V – the value of controlling voltage which doubles the oscillator frequency.

The output signal $u(t)$ of the phase detector PD is a response to the pair of signals $U_{\text{inp}}(\omega t)$ and $U_{\text{out}}\big(\theta(t)\big)$, where ω denotes the frequency of the input signal. Various equations of PD are considered in Ref. [9], [28]

9

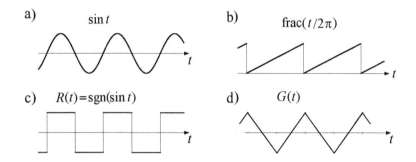

a) $\sin t$

b) $\mathrm{frac}(t/2\pi)$

c) $R(t) = \mathrm{sgn}(\sin t)$

d) $G(t)$

Fig. 2.2   Examples of periodic signals $U_{\mathrm{inp}}(t)$ or $U_{\mathrm{out}}(t)$.

and [33], for example

$$u(t) = U_{\mathrm{inp}}(\omega t) \cdot U_{\mathrm{out}}\big(\theta(t)\big). \tag{2.2}$$

For $U_{\mathrm{inp}}(\omega t) = 2AV \sin \omega t$ and for $U_{\mathrm{out}}(\theta) = \cos \theta$, the equation of the PLL system (see Fig.2.1) is

$$\frac{d\theta}{dt} = \Omega \left(1 + 2A \sin \omega t \, \cos \theta\right). \tag{2.3}$$

It is convenient to introduce a *phase difference* $\varphi(t) = \theta(t) - \omega t$. Then, the output signal of PLL is $\cos\big(\omega t + \varphi(t)\big)$, and the equation (2.3) is equivalent to

$$\frac{d\varphi}{dt} = A\Omega\big(\Delta - \sin \varphi + \sin(2\omega t + \varphi)\big), \tag{2.4}$$

where

$$\Delta = \frac{\Omega - \omega}{A\Omega}. \tag{2.5}$$

Similarly, if $U_{\mathrm{inp}}(\omega t) = 2AV R(\omega t)$ and $U_{\mathrm{out}}(\theta) = R(\theta)$, where $R(\tau) = \mathrm{sgn}(\sin \tau)$ is of rectangular waveform (see Fig.2.2c), then the equation of the PLL system is given by

$$\frac{d\theta}{dt} = \Omega\big(1 + 2AR(\omega t)\,R(\theta)\big), \tag{2.6}$$

or equivalently, by

$$\frac{d\varphi}{dt} = A\Omega\big(\Delta + 2R(\omega t)\,R(\varphi + \omega t)\big). \tag{2.7}$$

In alternative models of the PLL system, the high-frequency components are neglected in (2.4) or (2.7). The right-hand sides of these equations are replaced by their time-mean values. Then, we get the autonomous averaged equation

$$\frac{d\varphi}{dt} = A\Omega\left(\Delta - \sin\varphi\right), \tag{2.8}$$

instead of the equation (2.4), or

$$\frac{d\varphi}{dt} = A\Omega\left(\Delta + 2G(\varphi)\right), \tag{2.9}$$

instead of the equation (2.7), where $G(\varphi)$ is 2π-periodic triangular waveform function (see Fig.2.2d)

$$G(\varphi) = \frac{1}{2\pi} \int_{-\pi}^{\pi} R(\tau)R(\varphi + \tau)d\tau = 1 - \frac{2|\varphi|}{\pi} \quad \text{for} \quad -\pi \le \varphi \le \pi. \tag{2.10}$$

Without loss of generality we have assumed that the gain of a phase detector and the amplitude of an output signal are equal to one. In the real systems the parameter A depends also on these two quantities.

Let us consider one example more. Let the input signal be a periodic sequence of short impulses at the moments $t_n = 2\pi n/\omega$. Mathematical model of such a signal is so-called "comb function" $AV\delta_C(\omega t)$, where

$$\delta_C(\omega t) = 2\pi \sum_{n=-\infty}^{+\infty} \delta(\omega t - 2\pi n) \tag{2.11}$$

and $\delta(t)$ is Dirac's delta function.

Equation of PLL system takes the form

$$\frac{d\theta}{dt} = \Omega\left(1 + A\delta_C(\omega t)U_{\text{out}}(\theta)\right), \tag{2.12}$$

or

$$\frac{d\varphi}{dt} = A\Omega\left(\Delta + \delta_C(\omega t)U_{\text{out}}(\varphi + \omega t)\right). \tag{2.13}$$

If we neglect high-frequency components, i.e. if we replace the right-hand side of (2.13) by its mean value (with respect to t for a fixed φ), we obtain the autonomous averaged equation

$$\frac{d\varphi}{dt} = A\Omega\left(\Delta + U_{\text{out}}(\varphi)\right). \tag{2.14}$$

On the other hand, integrating the equation (2.12) in the interval (t_n, t_{n+1}), we get a recurrence equation of discrete-time system

$$\theta(t_{n+1}) = \theta(t_n) + 2\pi\frac{\Omega}{\omega}\left(1 + AU_{\text{out}}\left(\theta(t_n)\right)\right), \quad \omega t_n = 2\pi n, \tag{2.15}$$

where $\theta(t_n)$ is the left-hand side limit of the function $\theta(t)$ at the point t_n (the function $\theta(t)$ is discontinuous at the points t_n).

Both equations (2.14) and (2.15) are different approximate models of a real one-dimensional PLL system.

Averaged equations (2.8), (2.9), (2.14) are more simple then (2.4), (2.7), (2.13) and their solutions approximate the solutions of original equations, provided that ω is a large number. Averaged equations will be investigated in the next section. Discrete-time systems, similar to (2.15) will be discussed in Chapter 4.

Equation (2.8) is called *Adler's equation* (see Ref. [1]) and it is usually investigated in technical literature.

2.2   The averaged equation

The signals U_{inp} and U_{out} can be of different shapes. Similarly, the phase detector not always is described by the equation (2.2). However, in all these cases the right-hand side of averaged equation is 2π-periodic function of phase difference. Therefore it is reasonable to investigate properties of solutions of a more general differential equation

$$\frac{dx}{d\tau} = F(x), \tag{2.16}$$

where F is a bounded 2π-periodic function. We assume that for every initial value, the equation has exactly one solution (e.g. the function F satisfies the Lipschitz condition).

2.2.1   *Basic properties of solutions*

It is evident that if a function $x = \varphi(\tau)$ satisfies (2.16), then for every τ_0, the function $x = \varphi(\tau + \tau_0)$ also satisfies this equation.

Let $x = \varphi(\tau; p)$ denote the solution of equation (2.16) satisfying the initial condition $\varphi(0; p) = p$. As F is bounded, the solution $x = \varphi(\tau; p)$ exists for $\tau \in (-\infty, +\infty)$.

Theorem 2.1. *Let p be a fixed initial value. The solution $x = \varphi(\tau; p)$ of equation (2.16) has the following properties:*

(a) *if $F(p) = 0$, then $\varphi(\tau; p) \equiv p$ is the constant function,*

(b) *if $F(p) > 0$, then $\varphi(\tau; p)$ is an increasing function of τ,*

(c) *if $F(p) < 0$, then $\varphi(\tau; p)$ is a decreasing function of τ,*

(d) *if* $F(p) \neq 0$, *then* $x = \varphi(\tau; p)$ *has the inverse function*

$$\tau = \varphi^{-1}(x; p), \quad \text{where} \quad \varphi^{-1}(x; p) = \int_p^x \frac{d\xi}{F(\xi)}, \qquad (2.17)$$

(e) *let* $F(a) = F(b) = 0$, *where* $-\infty < a < b < +\infty$, *and* $a < p < b$. *If* $F(x) > 0$ *for* $a < x < b$, *then* $\varphi(\tau; p)$ *is an increasing function, and* $\varphi(-\infty; p) = a$, $\varphi(+\infty; p) = b$. *If* $F(x) < 0$ *for* $a < x < b$, *then* $\varphi(\tau; p)$ *is a decreasing function, and* $\varphi(-\infty; p) = b$, $\varphi(+\infty; p) = a$.

Proof. For every initial condition, the equation (2.16) has exactly one solution. Therefore properties (a), (b), (c), (e) follow immediately from (2.16). The function $\tau = \varphi^{-1}(x; p)$, which is inverse to the solution $x = \varphi(\tau; p)$ of equation (2.16), satisfies the differential equation $\frac{d\tau}{dx} = \frac{1}{F(x)}$, and this leads to the property (d). $\qquad\qquad\qquad\qquad\qquad\qquad\qquad \Box$

For a periodic function F some additional properties of the solutions $x = \varphi(\tau)$ of the equation

$$\frac{dx}{d\tau} = F(x), \quad \text{where} \quad F(x + 2\pi) = F(x) \qquad (2.18)$$

will be formulated.

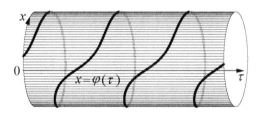

Fig. 2.3   Graph of a solution of Eq. (2.18) on the cylinder.

It is evident that if a function $x = \varphi(\tau)$ satisfies the equation (2.18), then for every integer n, the function $x = \varphi(\tau) + 2\pi n$ also satisfies (2.18). All these solutions (for all integers n) have the same graph on the cylinder $R \times S$ (Cartesian product of line R and circle S), where $x \bmod 2\pi$ is a cyclic variable (see Fig.2.3). Therefore the cylinder is a more adequate surface for illustrating properties of solutions of equation (2.18).

Theorem 2.2. *If* $F(x) > 0$ *or* $F(x) < 0$ *for all* x, *then every solution of equation* (2.18) *is of the form*

$$x = \varphi(\tau - \tau_*), \qquad \varphi(\tau) = s\tau + \beta(s\tau), \qquad (2.19)$$

where τ_* is an arbitrary constant, $\beta(\psi) = \beta(\psi + 2\pi)$ is a periodic function and the parameter s (the frequency of the periodic component) is given by

$$s = \left(\frac{1}{2\pi} \int_0^{2\pi} \frac{d\xi}{F(\xi)} \right)^{-1}. \tag{2.20}$$

Moreover, the amplitude A_0 of the periodic component is

$$A_0 \overset{\text{def}}{=} \max_{\psi_0, \psi_1} \frac{\beta(\psi_1) - \beta(\psi_0)}{2} = \max_{x_0, x_1} \frac{1}{2} \int_{x_0}^{x_1} \left(1 - \frac{s}{F(\xi)} \right) d\xi. \tag{2.21}$$

Proof. Let us assume that $F(x) > 0$ (if $F(x) < 0$ then the proof is similar). Every solution $x = \varphi(\tau)$ of equation (2.18) is a monotonic function increasing from $-\infty$ to $+\infty$. For a fixed τ_0, let us denote by $T = T(\tau_0)$ such a number, that $\varphi\big(\tau_0 + T(\tau_0)\big) = \varphi(\tau_0) + 2\pi$. For every τ_0, τ_1 we have

$$\int_{\varphi(\tau_0)}^{\varphi(\tau_1)} \frac{d\xi}{F(\xi)} = \tau_1 - \tau_0. \tag{2.22}$$

If $\tau_1 = \tau_0 + T(\tau_0)$, then

$$T(\tau_0) = \int_{\varphi(\tau_0)}^{\varphi(\tau_0)+2\pi} \frac{d\xi}{F(\xi)} = \int_0^{2\pi} \frac{d\xi}{F(\xi)}. \tag{2.23}$$

The last equality is a consequence of periodicity of $F(x)$. Hence, $T(\tau_0)$ does not depend on τ_0.

For $s = \frac{2\pi}{T}$, the function $\varphi(\tau) - s\tau$ is periodic of the period T:

$$\big(\varphi(\tau + T) - s(\tau + T)\big) - \big(\varphi(\tau) - s\tau\big) = \varphi(\tau + T) - \varphi(\tau) - sT = 0,$$

and $\beta(s\tau) \overset{\text{def}}{=} \varphi(\tau) - s\tau$ is a 2π-periodic function with respect to $\psi = s\tau$. Let τ_0, τ_1 be two arbitrary points. Let us denote

$$x_0 = \varphi(\tau_0) = s\tau_0 + \beta(s\tau_0), \qquad x_1 = \varphi(\tau_1) = s\tau_1 + \beta(s\tau_1).$$

Using (2.22), we get the equality

$$\beta(s\tau_1) - \beta(s\tau_0) = x_1 - x_0 - s(\tau_1 - \tau_0) = \int_{x_0}^{x_1} \left(1 - \frac{s}{F(\xi)} \right) d\xi,$$

from which we conclude (2.21). $\qquad\qquad\qquad\qquad\qquad\qquad\qquad\square$

2.2.2 Application to Adler's equation

Theorems 2.1 and 2.2 will be applied to describe the solutions of the Adler's equation (2.8)

If $|\Delta| < 1$, then there exist two ($mod\ 2\pi$) constant solutions:

$$\varphi_s = \arcsin \Delta, \qquad \varphi_u = \pi - \arcsin \Delta \qquad (2.24)$$

and two families of monotonic solutions: increasing solutions with values in the interval $(\varphi_u - 2\pi, \varphi_s)$ and decreasing solutions with values in the interval (φ_s, φ_u), as shown in Fig.2.4a. The constant solution $\varphi_s\ mod\ 2\pi$ is asymptotically stable, and almost all initial values $\varphi(0)$ (except $\varphi(0) = \varphi_u\ mod\ 2\pi$) belong to its basin of attraction. The output signal of PLL is $\cos\big(\omega t + \varphi(t)\big)$, and it tends to $\cos(\omega t + \varphi_s)$ as $t \to \infty$.

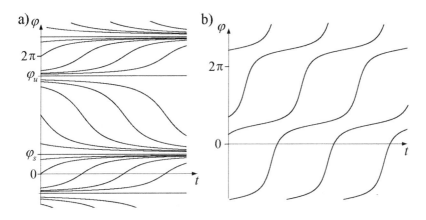

Fig. 2.4   Graphs of a family of solutions of Eq. (2.8).

If $|\Delta| > 1$, then all solutions of equation (2.8) are increasing (for $\Delta > 1$) or decreasing (for $\Delta < -1$). The solutions $\varphi(t) = st + \beta(st)$ are stable but not asymptotically stable. According to Theorem 2.2 the frequency and amplitude of the periodic component are

$$s = A\Omega\sqrt{\Delta^2 - 1}\ \text{sgn}\,\Delta, \qquad A_0 = 2\arcsin\left(|\Delta| - \sqrt{\Delta^2 - 1}\right).$$

Graphs of a family of solutions are shown in Fig.2.4b. The amplitude A_0 and normalized frequency $|s|/A\Omega$ of the periodic component $\beta(st)$ versus Δ are shown in Fig.2.5. The output signal of the PLL system is an almost periodic function

$$\cos\theta(t) = \cos\big(\omega t + s(t - t_*) + \beta(st - st_*)\big), \qquad (2.25)$$

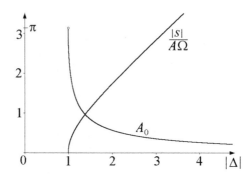

Fig. 2.5   The amplitude A_0 and normalized frequency $|s|/A\Omega$ of the periodic component of solutions of Eq. (2.8) versus $|\Delta|$.

where t_* is an arbitrary constant dependent on an initial value.

Let us introduce the following definition:

Definition 2.1. Let $p(\tau) = p(\tau + 2\pi)$ be a periodic function, and let $\theta(t)$ be an increasing function. The number (if it exists)

$$\omega_c = \lim_{t\to\infty} \frac{\theta(t)}{t} \qquad (2.26)$$

is said to be the center-frequency of the signal $p\big(\theta(t)\big)$.

The center-frequency of the output signal of PLL is equal to $(\omega + s)$ for $|\Delta| > 1$, and is equal to ω for $|\Delta| < 1$. Consequently, the ratio $\frac{\omega_c}{\omega}$, as a function of $\frac{\Omega}{\omega}$, is given by the formula

$$\frac{\omega_c}{\omega} = \begin{cases} 1 + \sqrt{\left(\frac{\Omega}{\omega} - 1\right)^2 - A^2 \left(\frac{\Omega}{\omega}\right)^2}\, \operatorname{sgn}(\Omega - \omega) & \text{for} \quad A < \left|1 - \frac{\omega}{\Omega}\right| \\ 1 & \text{for} \quad A > \left|1 - \frac{\omega}{\Omega}\right| \end{cases}$$

and its graph is shown in Fig.2.6.

The almost periodic function (2.25) can be expanded into the following Fourier series:

$$\cos\theta(t) = \operatorname{Re}\left(\mathrm{e}^{j\theta(t)}\right) = \operatorname{Re}\left(\sum_{k=-\infty}^{+\infty} c_k \mathrm{e}^{j\nu_k t}\right), \quad \text{where } \nu_k = \omega + ks. \quad (2.27)$$

The sequence of the Fourier coefficients $c_k = c_k(\nu_k)$ is called the *spectrum* of the signal (2.25). The spectrum depends on the parameters A and $\frac{\Omega}{\omega}$. Examples of the amplitude-spectra $|c_k(\nu_k)|$ are shown in Fig.2.7 for a fixed $\frac{\Omega}{\omega}$ and for three values of the parameter A.

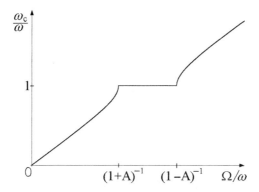

Fig. 2.6   The center-frequency ω_c versus the quiescent frequency Ω of the VCO.

It is possible to give an analytic formula for the solution of equation (2.8). The function inverse to the solution of (2.8) is

$$\frac{1}{A\Omega} \int^{\varphi} \frac{d\varphi}{\Delta - \sin\varphi} \equiv \frac{2}{A\Omega\sqrt{\Delta^2 - 1}} \arctan\left(\sqrt{\frac{\Delta + 1}{\Delta - 1}} \tan\frac{2\varphi - \pi}{4}\right) + const.$$

and therefore the solution takes the form

$$\varphi(t) = 2\arctan\left(\sqrt{\frac{\Delta - 1}{\Delta + 1}} \tan\frac{A\Omega\sqrt{\Delta^2 - 1}(t - t_*)}{2}\right) + \frac{\pi}{2},$$

where t_* is an arbitrary constant.

However, this formula is too complicated (the function *arctan* is multi-valued, the formula must be reduced to the real-valued function for $|\Delta| < 1$,

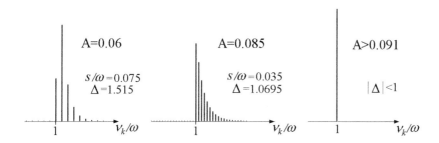

Fig. 2.7   Amplitude-spectra of the output signal $\cos\theta(\tau)$ of PLL for $\Omega/\omega = 1.1$ and for three values of the parameter A. The distance between two adjacent frequencies of spectra is equal to s/ω.

the limit value must be calculated for $|\Delta| = 1$), and it is better to obtain qualitative properties of solutions as consequences of Theorems 2.1 and 2.2.

2.3   Solutions of the basic frequency

In the previous section the high-frequency component of the signal $u(t)$ was neglected. However it has considerable influence on the solutions of PLL system in a long time-interval (e.g an interval equal to hundreds periods of the input signal). In particular the constant solution of averaged equation is replaced by a periodic solution. If it is asymptotically stable then the output signal of PLL is synchronized with basic frequency of the input signal. Such periodic solutions are investigated in this section. In this order we consider a general differential equation

$$\frac{dx}{d\tau} = F(x, \tau), \quad \text{where} \quad F(x, \tau + 2\pi) = F(x, \tau), \tag{2.28}$$

which can be used to investigate the PLL systems with various sinusoidal or non-sinusoidal signals U_{inp}, U_{out}.

Let us suppose that for every initial value the equation has exactly one solution which continuously depends on the initial value.

2.3.1   *The Poincaré mapping*

Let $x = \varphi(\tau; p)$ denote the solution of equation (2.28) with an initial value $\varphi(0; p) = p$. The notion given below is fundamental for the theory of equation (2.28), (see Refs. [48], [50]).

Definition 2.2. The function f which maps an initial value p onto the value

$$f(p) = \varphi(2\pi; p) \tag{2.29}$$

of the solution at the moment $\tau = 2\pi$ is called the Poincaré mapping.

Let f^n denote the composition of f with itself n-times $f^n(p) = f\bigl(f^{n-1}(p)\bigr)$. The function f^0 is the identity, and f^{-1} is the inverse of f.

As $F(x, \tau)$ is a periodic function with respect to τ, we have $\varphi(\tau + 2\pi; p) = \varphi\bigl(\tau; f(p)\bigr)$ and, consequently,

$$f^n(p) = \varphi(2\pi n; p). \tag{2.30}$$

Proposition 2.1. *The Poincaré mapping is an increasing function. If there exists a continuous derivative $F'_x(x, \tau)$ then*

$$f'(p) = \exp\left(\int_0^{2\pi} F'_x\big(\varphi(\tau; p), \tau\big)d\tau \right). \tag{2.31}$$

Proof. The function $z(\tau, p) = \dfrac{\partial \varphi(\tau; p)}{\partial p}$ satisfies the linear differential equation $\dfrac{dz}{d\tau} = F'_x\big(\varphi(\tau; p), \tau\big)z$ and the initial condition $z(0, p) = 1$. The formula (2.31) is a consequence of the equality $f'(p) = z(2\pi, p)$. $\quad\square$

For a fixed p the sequence

$$\ldots f^{-n}(p), \ldots, f^{-2}(p),\ f^{-1}(p),\ p,\ f(p),\ f^2(p), \ldots, f^n(p), \ldots \tag{2.32}$$

is monotonic: increasing for $p < f(p)$, decreasing for $p > f(p)$, and constant for $p = f(p)$. The solution $x = \varphi(\tau; p_0)$ of equation (2.28) is a 2π-periodic function if and only if the initial value p_0 satisfies the equation $p = f(p)$. This solution is asymptotically stable for $f'(p_0) < 1$ and unstable for $f'(p_0) > 1$.

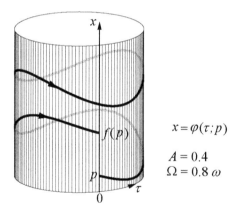

$$x = \varphi(\tau; p)$$
$$A = 0.4$$
$$\Omega = 0.8\,\omega$$

Fig. 2.8   Graphs of two solutions of Eq. (2.4) on the cylinder (one of them is periodic and asymptotically stable).

It is evident that if the function $x = \varphi(\tau)$ satisfies the equation (2.28), then for every integer n, the function $x = \varphi(\tau + 2\pi n)$ also satisfies (2.28). All these solutions (for all integers n) have the same graph (see Fig.2.8) on the cylinder $S \times R$, where $\tau \bmod 2\pi$ is a cyclic variable. The graph of a periodic solution is a closed curve which surrounds the cylinder.

A family of solutions $\varphi(\tau; p)$ (where $\tau = \omega t$ is the phase of the input signal) of the equation (2.4) for several initial values p and the Poincaré mapping $f(p) = \varphi(2\pi; p)$ are shown in Fig.2.9 for parameters $\Omega = 1.2\,\omega$ and $A = 0.25$. Graphs of two selected solutions on the cylinder $S \times R$ are shown in Fig.2.8 for other indicated values of the parameters.

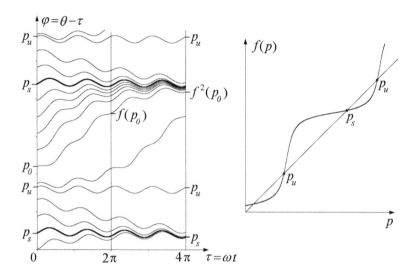

Fig. 2.9   The family of solutions of Eq. (2.4) and the graph of the Poincaré mapping.

2.3.2   *Periodic solutions*

The existence of a periodic solution of equation (2.28) follows from properties of the sequence (2.32).

Proposition 2.2. *The equation* (2.28) *has a periodic solution if and only if it has an upper-bounded or a lower-bounded solution.*

Proof.   If $x = \varphi(\tau, p)$ is an upper-bounded or a lower-bounded solution, then the monotonic sequence (2.32) converges to a point p_0 as n tends to $+\infty$ or $-\infty$. The point p_0 satisfies the equality $p_0 = f(p_0)$, and $x = \varphi(\tau, p_0)$ is the periodic solution. □

Let p_a, p_b be two fixed numbers. Conditions of the existence of a periodic solutions of equation (2.28) in the domain

$$D = \{\tau, x: \quad \tau \in (-\infty, +\infty), \quad p_a < x < p_b\} \qquad (2.33)$$

will be given.

Theorem 2.3. *The equation* (2.28) *has the following properties:*

(a) *If $F(x, \tau) \neq 0$ in every point of D, then periodic solutions do not exist in D.*

(b) *If $F(p_a, \tau) > 0$ and $F(p_b, \tau) < 0$ (or if $F(p_a, \tau) < 0$ and $F(p_b, \tau) > 0$) for $\tau \in [0, 2\pi]$, then a periodic solution exists in D.*

(c) *If there exists a continuous derivative $F'_x(x, \tau)$, and $F'_x(x, \tau) \neq 0$ in every point of D, then at most one periodic solution exists in D. It is stable for $F'_x(x, \tau) < 0$, and unstable for $F'_x(x, \tau) > 0$.*

Proof. (a) Let a periodic solution $x = \varphi(\tau)$ of equation (2.28) exist in D. Its derivative $\frac{d\varphi(\tau)}{d\tau} = F\big(\varphi(\tau), \tau\big)$ changes the sign. This contradicts the assumption.

(b) If $F(p_a, \tau) > 0$, then the solution $x = \varphi(\tau; p_a)$ is an increasing function in a neighborhood of the point $(0, p_a)$ and its graph does not intersect the line $x = p_a$ for $\tau > 0$. Consequently, $f(p_a) > p_a$. Similarly, if $F(p_b, \tau) < 0$ then $f(p_b) < p_b$. Since the Poincaré mapping is continuous, there exists $p_0 \in (p_a, p_b)$ such that $p_0 = f(p_0)$. Finally, $x = \varphi(\tau; p_0)$ is the periodic solution.

(c) Let two periodic solutions $x = \varphi_1(\tau)$ and $x = \varphi_2(\tau)$ exist in D. The periodic function $z(\tau) = \varphi_1(\tau) - \varphi_2(\tau)$ is not equal to zero in each point. It satisfies the differential equation

$$\frac{dz(\tau)}{d\tau} = F\big(\varphi_1(\tau), \tau\big) - F\big(\varphi_2(\tau), \tau\big) = F'_x\big(\tilde{x}(\tau), \tau\big) z(\tau),$$

where $\tilde{x}$ is a periodic function such that $\varphi_2(\tau) \leq \tilde{x}(\tau) \leq \varphi_1(\tau)$. After integration we get

$$z(2\pi) - z(0) = \int_0^{2\pi} F'_x\big(\tilde{x}(\tau), \tau\big) z(\tau) d\tau.$$

The left-hand side of this equality is zero, but right-hand side is not equal to zero. Consequently, the existence of two periodic solutions is not possible. The stability or non-stability of a periodic solution follows from (2.31). $\square$

2.3.3 *Asymptotic formulae for periodic solutions*

Let us consider the differential equation

$$\frac{dx}{d\tau} = \kappa F(x, \tau), \quad \text{where } F(x, \tau + 2\pi) = F(x, \tau), \quad (2.34)$$

where F is of class C^2, and κ is a positive parameter. If κ is small then we can use the averaging method. However, for the periodic solution a more precise formula can be given.

Theorem 2.4. *Let x_∞ be a number such that*

$$\frac{1}{2\pi} \int_0^{2\pi} F(x_\infty, \tau) d\tau = 0 \quad and \quad m \overset{\text{def}}{=} \frac{1}{2\pi} \int_0^{2\pi} F'_x(x_\infty, \tau) d\tau \neq 0. \quad (2.35)$$

There exists κ^* such that for all $0 < \kappa < \kappa^*$, in a neighborhood of x_∞, the equation (2.34) has exactly one 2π-periodic solution $x = \varphi_\kappa(\tau)$ continuously dependent on κ and $\sup_\tau |\varphi_\kappa(\tau) - x_\infty| \to 0$ as $\kappa \to 0$.

This solution is of the form

$$\varphi_\kappa(\tau) = x_\infty + \kappa \varphi_1(\tau) + O\left(\kappa^2\right), \quad (2.36)$$

where the function φ_1 is completely defined by

$$\frac{d\varphi_1(\tau)}{d\tau} = F(x_\infty, \tau), \quad and \quad \int_0^{2\pi} F'_x(x_\infty, \tau) \varphi_1(\tau) d\tau = 0. \quad (2.37)$$

The solution is stable for $m < 0$ and unstable for $m > 0$.

Proof.   A solution of equation (2.34) depends on the initial value p and on the parameter κ. It satisfies the integral equation

$$x(\tau; p, \kappa) = p + \kappa \int_0^\tau F\big(x(t; p, \kappa), t\big) dt.$$

The solution is periodic if and only if the mean value of the periodic function $F\big(x(t; p, \kappa), t\big)$ is equal to zero. Let us introduce the function

$$G(p, \kappa) = \frac{1}{2\pi} \int_0^{2\pi} F\big(x(\tau; p, \kappa), \tau\big) d\tau$$

which is continuous and has the continuous derivative $G'_p(p, \kappa)$. It is easy to check that $G(x_\infty, 0) = 0$ and $G'_p(x_\infty, 0) = m$. By the implicit function theorem (see Ref. [50], [24]), in a neighborhood of zero, there exists exactly one solution $p = \psi(\kappa)$ of the equation $G(p, \kappa) = 0$. The function $p = \psi(\kappa)$ is continuous in a neighborhood of zero, and $\psi(0) = x_\infty$. The equation (2.34) with the initial value $p = \psi(\kappa)$ has a periodic solution.

The formula (2.36) can be obtained in the following way. The solution is assumed in the form $x(\tau) = \varphi_0(\tau) + \kappa \varphi_1(\tau) + \kappa^2 \varphi_2(\tau) + \ldots$, where each function $\varphi_n(\tau)$ is periodic. Replacing in (2.34) x by this formal series and comparing terms of the same power of coefficient κ, we obtain a system of equations for φ_n. This system can be successively solved.

According to (2.31) the derivative of the Poincaré mapping at the point $p = \psi(\kappa)$ is equal to $f'(p) = \exp\big(2\pi m + O(\kappa)\big)$. Consequently, for sufficiently small values of κ, the periodic solution is stable for $m < 0$ and unstable for $m > 0$. □

The next theorem can be proved in a similar way.

Theorem 2.5. *Let $x_0(\tau)$ denote a 2π-periodic function which satisfies conditions*

$$F\big(x_0(\tau),\tau\big) = 0 \quad and \quad m(\tau) \overset{\text{def}}{=} F'_x\big(x_0(\tau),\tau\big) \neq 0. \tag{2.38}$$

There exists κ^* such that for all $\kappa > \kappa^*$, in a neighborhood of $x_0(\tau)$, the equation (2.34) has exactly one 2π-periodic solution $x = \varphi_\kappa(\tau)$ continuously dependent on κ and $\sup_\tau |\varphi_\kappa(\tau) - x_0(\tau)| \to 0$ as $\kappa \to \infty$.

This solution is of the form

$$\varphi_\kappa(\tau) = x_0(\tau) - \frac{1}{\kappa}\frac{F'_\tau\big(x_0(\tau),\tau\big)}{\big(m(\tau)\big)^2} + O\left(\kappa^{-2}\right). \tag{2.39}$$

The solution is stable for $m(\tau) < 0$ and unstable for $m(\tau) > 0$.

The proof is omitted.

2.3.4 Conclusions for the PLL equation

The asymptotic formulae for periodic solutions of equation (2.4) will be given for small values of the parameter A.

Proposition 2.3. *If $|\Delta| < 1$ and if A is sufficiently small, then there exist exactly two $(\mathrm{mod}\, 2\pi)$ periodic solutions of equation (2.4):*
the stable solution

$$\varphi_{\text{stab}}(t) = \varphi_\infty - A\frac{\Omega}{\omega}\left(\frac{1}{4\cos\varphi_\infty} + \frac{1}{2}\cos(2\omega t + \varphi_\infty)\right) + O\left(A^2\right) \tag{2.40}$$

and the unstable solution

$$\varphi_{\text{unst}}(t) = \pi - \varphi_\infty + A\frac{\Omega}{\omega}\left(\frac{1}{4\cos\varphi_\infty} + \frac{1}{2}\cos(2\omega t - \varphi_\infty)\right) + O\left(A^2\right), \tag{2.41}$$

where $\varphi_\infty = \arcsin\Delta \in (-\frac{\pi}{2},+\frac{\pi}{2})$.

Proof. According to Theorem 2.4 (where $x = \varphi$, $\tau = \omega t$ and $\kappa = A\Omega/\omega$) we have $\sin\varphi_\infty = \Delta$ and $m = -\cos\varphi_\infty$. From (2.37) it follows that

$$\varphi_1(\tau) = -\frac{1}{4\cos\varphi_\infty} - \frac{1}{2}\cos(2\tau + \varphi_\infty).$$

Equation $\sin\varphi_\infty = \Delta$ has two solutions: $\varphi_{\infty,1} = \arcsin\Delta$ with $m < 0$ and $\varphi_{\infty,2} = \pi - \arcsin\Delta$ with $m > 0$. Consequently, two $(\mathrm{mod}\ 2\pi)$ periodic solutions exist. □

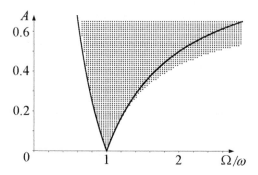

Fig. 2.10   The domain of existence of an asymptotically stable $(2\pi/\omega)$-periodic solution of Eq. (2.4).

If the equation (2.4) has a stable 2π-periodic solution, then the steady--state output signal

$$U_{\text{out}}\big(\theta(t)\big) = \cos\big(\omega t + \varphi_{\text{stab}}(t)\big) \qquad (2.42)$$

is periodic of the same period $2\pi/\omega$ as the input signal, what means a synchronization of the PLL. The domain of existence of such solutions is shown in Fig.2.10 in the plane of parameters $\frac{\Omega}{\omega}, A$. If A is small, then this domain is approximately described by $A > |1 - \frac{\omega}{\Omega}|$ (solid line in the picture).

2.4   Differential equation on the torus

Fundamental notions and theorems about the differential equation

$$\frac{dx}{d\tau} = F(x,\tau), \quad \text{where} \quad F(x+2\pi,\tau) = F(x,\tau) = F(x,\tau+2\pi), \qquad (2.43)$$

will be given in this section. They are important for the problems of synchronization several types of physical oscillators, particularly the PLL systems described by the equations (2.3) or (2.6).

2.4.1   Trajectories on the torus

If $x = \varphi(\tau)$ is a solution of (2.43), then for every integers n, m, the functions

$$\varphi(\tau + 2\pi m) + 2\pi n \qquad (2.44)$$

also satisfy the equation (2.43). Consequently, the graph of a family of solutions (see Fig.2.11) is double periodic with the rectangle of periodicity

$$D = \{\tau, x : \tau \in [0, 2\pi), x \in [0, 2\pi)\}. \tag{2.45}$$

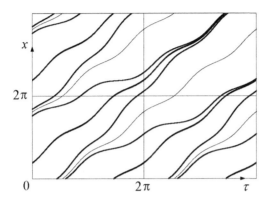

Fig. 2.11   The graph of a family of solutions of Eq.(2.43) in the τx-plane.

It is convenient to assume that both variables τ, x are cyclic *mod* 2π, and use the torus $S \times S$ as a manifold on which the graphs of solutions are pictured. Geometrical transformation of a rectangle of periodicity onto the torus is shown in Fig.2.12. The functions (2.44), for all integers n, m, have the same graph on the torus but an infinite number of graphs in the plane (see Ref. [2]).

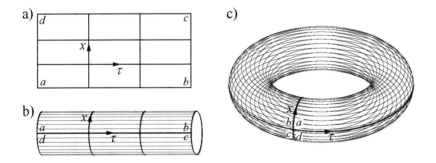

Fig. 2.12   Transformation of a rectangle onto the torus.

For a fixed number ρ, let us consider the equation $\frac{dx}{d\tau} = \rho$. It is a particular case of the equation (2.43). The graph of each solution $x(\tau) = x(0) + \rho\tau$ surrounds the torus and intersects τ-axis with the slope equal to ρ. If $\rho = \frac{n}{m}$ is a rational number, then the graph of each solution is such a closed line which surrounds the torus m-times in direction of τ-axis and n-times in direction of x-axis. If ρ is an irrational number, then the graph of each solution is not a closed line but it surrounds the torus an infinite number times and is uniformly dense on the torus (see Ref. [51]).

2.4.2 *Periodic points*

Let $f(p) = \varphi(2\pi; p)$ be the Poincaré mapping for the equation (2.43). Periodicity of the function $F(x, \tau)$ implies

$$f(p + 2\pi) = f(p) + 2\pi \tag{2.46}$$

and, consequently, $f^m(p + 2\pi) = f^m(p) + 2\pi$ for each integer m or, equivalently, the function $f^m(p) - p$ is 2π-periodic.

If $p \bmod 2\pi$ is a cyclic variable, then the function f is a continuous mapping of the circle onto itself.

Definition 2.3. If there exist two coprime natural numbers n, m and a point p_0 such that

$$f^m(p_0) = p_0 + 2\pi n, \tag{2.47}$$

then p_0 is called a periodic point of the type n/m of the map f.

If p_0 is a periodic point of the type n/m, then the solution $x = \varphi(\tau; p_0)$ of equation (2.43) satisfies the identity

$$\varphi(\tau + 2\pi m; p_0) = \varphi(\tau; p_0) + 2\pi n$$

and its graph is a closed curve which surrounds the torus m-times in direction of τ-axis, and n-times in direction of x-axis. This curve is called a *periodic trajectory* or *periodic orbit* of the type n/m.

If p_0 is a periodic point of the type n/m, then each of the points

$$p_0, \ f(p_0), \ f^2(p_0), \ldots, f^{m-1}(p_0) \quad (mod \, 2\pi) \tag{2.48}$$

of the circle S is the periodic point of the same type. The system of points (2.48) is also called a *periodic orbit* of the point p_0 (the orbit of the Poincaré mapping f). The graph of the solution $x = \varphi(\tau; p_0)$ of equation (2.43) intersects the circle $\tau = 0$ on the torus $S \times S$ at the points (2.48) (see Ref. [16]).

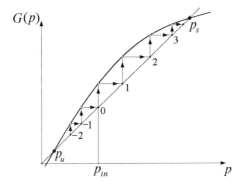

Fig. 2.13   Geometrical interpretation of the sequence (2.50). Integers $k = \dots - 2, -1,$
$0, 1, 2, 3\dots$ denote the points for which both coordinates are equal to $G^k(p_{in})$.

A periodic point p_0 of the type n/m is the fixed point of the map

$$G(p) \overset{\text{def}}{=} f^m(p) - 2\pi n. \tag{2.49}$$

Let f have a continuous derivative. The derivative of the increasing function G at the periodic point p_0 depends on the whole periodic orbit

$$G'(p_0) = f'(p_0) \cdot f'(p_1) \cdot \dots \cdot f'(p_{m-1}), \quad \text{where} \quad p_k = f^k(p_0).$$

Definition 2.4. A periodic point p_0 and corresponding periodic orbit is called

– stable   if $G'(p_0) < 1$,
– unstable if $G'(p_0) > 1$,
– neutral if $G'(p_0) = 1$.

In Fig.2.13 it is shown how for a given initial value p_{in}, the monotonic sequence

$$\dots, G^{-2}(p_{in}), \quad G^{-1}(p_{in}), \quad p_{in}, \quad G^1(p_{in}), \quad G^2(p_{in}), \dots \tag{2.50}$$

tends from an unstable fixed point p_u to a stable fixed point p_s.

2.4.3   Rotation number

The basic characteristic of solutions of the equation (2.43) is so called *rotation number*:

$$\rho = \lim_{m \to \infty} \frac{f^m(p)}{2\pi m}. \tag{2.51}$$

Theorem 2.6. *Let f be a continuous and nondecreasing function which satisfies the condition (2.46). Then,*

(a) *The limit (2.51) exists and does not depend on the point p.*

(b) *The map f has a periodic point p_0 if and only if $\rho = \frac{n}{m}$ (where n, m are coprime) is a rational number. Then the periodic point p_0 is of the type n/m and it satisfies the equality $f^m(p_0) = p_0 + 2\pi n$.*

(c) *If ρ is an irrational number and the derivative f' is a continuous and positive function of bounded variation, then the mapping f is topologically equivalent to the rotation of the circle through the angle $2\pi\rho$.*

For the proof of the theorem we refer the readers to Refs. [4], [57] or [44].

A similar theorem can be given for solutions of the equation (2.43). Its successive parts follow from the corresponding parts of Theorem 2.6.

Theorem 2.7. *Let $F(x, \tau)$ (the right hand side of (2.43)) be a continuous function of both variables and let it have the continuous partial derivative $F'_x(x, \tau)$. Then,*

(a) *For each solution $x = \varphi(\tau; p)$ of equation (2.43) there exists the limit*

$$\rho = \lim_{\tau \to \infty} \frac{\varphi(\tau; p)}{\tau} \qquad (2.52)$$

(the rotation number) and it does not depend on the initial value p.

(b) *The equation (2.43) has a closed trajectory on the torus if and only if $\rho = \frac{n}{m}$ is a rational number. Then there exists such a p_0 that the graph of the solution $x = \varphi(\tau; p_0)$ surrounds the torus m-times in direction of τ-axis and n-times in direction of x-axis.*

(c) *If ρ is an irrational number and F has the continuous partial derivative $F''_{xx}(x, \tau)$ of the second order then on the torus there exists such a system of coordinates τ', x' that the mapping $(\tau', x') \to (\tau, x)$ is a homeomorphism (one to one correspondence continuous in both directions) and in the new system of coordinates the equation (2.43) is of the form $\frac{dx'}{d\tau'} = \rho$.*

2.4.4   Rotation number as the function of a parameter

Physicists and engineers are specially interested in such periodic solutions of the equation (2.43) which are asymptotically stable and their type n/m does not change for small perturbation of amplitude or frequency of an input signal. Such solutions correspond to fractional synchronization and they are important for coherent transponders. Therefore, now we investigate how the rotation number depends on parameters.

Let us consider the differential equation

$$\frac{dx}{d\tau} = F(x, \tau; \mu), \quad \text{where} \quad F(x+2\pi, \tau; \mu) = F(x, \tau; \mu) = F(x, \tau+2\pi; \mu), \quad (2.53)$$

with a parameter μ. The solution $x = \varphi(\tau; p, \mu)$, the Poincaré mapping $f(p, \mu) = \varphi(2\pi; p, \mu)$ and the rotation number

$$\rho(\mu) = \lim_{\tau \to \infty} \frac{\varphi(\tau; p, \mu)}{\tau} = \lim_{m \to \infty} \frac{f^m(p, \mu)}{2\pi m} \quad (2.54)$$

depend on the parameter.

Theorem 2.8. *Let the function $f(p, \mu)$ be continuous with respect to both variables, nondecreasing with respect to p and*

$$f(p + 2\pi, \mu) = f(p, \mu) + 2\pi. \quad (2.55)$$

Then the rotation number $\rho(\mu)$ is a continuous function of the parameter μ.

If, moreover, $f(p, \mu)$ is a nondecreasing function of μ, then $\rho(\mu)$ is also nondecreasing.

Proof. Let the numbers p and μ_0 be arbitrarily fixed. For every $\varepsilon > 0$ and for a natural number $m > 2/\varepsilon$ there exists an integer n such that

$$2\pi(n - 1) < f^m(p, \mu_0) < 2\pi(n + 1).$$

As the function f is continuous for $\mu = \mu_0$, then there exists a number $\delta > 0$ such that

$$2\pi(n - 1) < f^m(p, \mu) < 2\pi(n + 1) \quad \text{for} \quad |\mu - \mu_0| < \delta.$$

By the definition of the rotation number, we have

$$\frac{n - 1}{m} \le \rho(\mu_0) \le \frac{n + 1}{m} \quad \text{and} \quad \frac{n - 1}{m} \le \rho(\mu) \le \frac{n + 1}{m}.$$

From the above it follows that

$$|\rho(\mu_0) - \rho(\mu)| \le \frac{2}{m} < \varepsilon \quad \text{for} \quad |\mu_0 - \mu| < \delta, \quad (2.56)$$

and the function $\rho(\mu)$ is continuous at the point μ_0.

If $f(p, \mu_1) \le f(p, \mu_2)$ for $\mu_1 < \mu_2$ and for every p, then also $f^m(p, \mu_1) \le f^m(p, \mu_2)$ and by the definition of the rotation number $\rho(\mu_1) \le \rho(\mu_2)$. This finishes the proof. $\square$

Remark. Generally, $\rho(\mu)$ has no bounded derivative even if $f(p, \mu)$ is an analytic function.

Theorem 2.9. *Let the right-hand side of the equation (2.53) be a continuous function which satisfies the Lipschitz condition with respect to x, and let it be nondecreasing with respect to μ. Then the rotation number $\rho(\mu)$ is a continuous and nondecreasing function.*

Proof.   It is sufficient to check that the Poincaré mapping $f(p, \mu)$ satisfies the assumptions of Theorem 2.8.

For every initial value p, the equation (2.53) has exactly one solution $x = \varphi(\tau; p, \mu)$ which is a continuous function of the values (τ, p, μ). The Poincaré mapping is also a continuous function of both variables. It is an increasing function of p and $f(p + 2\pi, \mu) = f(p, \mu) + 2\pi$. We will prove that $\varphi(\tau; p, \mu)$ is a nondecreasing function of μ.

Let $x_1 = \varphi(\tau; p, \mu_1)$ and $x_2 = \varphi(\tau; p, \mu_2)$ denote solutions of equation (2.53) for two values of μ, where $\mu_1 < \mu_2$. We have

$$\frac{d(x_2 - x_1)}{d\tau} = \big(F(x_2, \tau; \mu_2) - F(x_1, \tau; \mu_2)\big) + \big(F(x_1, \tau; \mu_2) - F(x_1, \tau; \mu_1)\big) \quad (2.57)$$

Let us define two functions

$$B(\tau) = \frac{F(x_2, \tau; \mu_2) - F(x_1, \tau; \mu_2)}{x_2 - x_1}, \qquad \varepsilon(\tau) = F(x_1, \tau; \mu_2) - F(x_1, \tau; \mu_1).$$

The function $B(\tau)$ is bounded, and $\varepsilon(\tau) \geq 0$. The function $y = x_2 - x_1$ satisfies differential equation $\frac{dy}{d\tau} = B(\tau)y + \varepsilon(\tau)$ with the initial value $y(0) = = 0$ and it has nonnegative solution

$$y(\tau) = \int_0^\tau N(\tau, t)\varepsilon(t)dt, \quad \text{where} \quad N(\tau, t) = \exp\left(\int_t^\tau B(\xi)d\xi\right) \geq 0.$$

So, the solution $x = \varphi(\tau; p, \mu)$ of equation (2.53) is a nondecreasing function of μ and the Poincaré mapping $f(p, \mu)$ has the same property . All assumptions of Theorem 2.8 are satisfied. This finishes the proof.   □

2.5   Fractional synchronization

Rotation number $\rho(\mu)$, where $\mu = \frac{\Omega}{\omega}$, is a fundamental characteristic of synchronized oscillators (with self-frequency Ω and input-frequency ω). For PLL systems $\rho(\mu)$ is a continuous, nondecreasing and piecewise constant function. For explanation of strange shape of this function it is necessary to introduce two notions: Cantor's step-function (devil's staircase) and so called *T-property* (a property, which is typical).

2.5.1   *Devil's staircase*

Definition 2.5. A continuous and nondecreasing function $g(x)$ which maps the interval $[x_1, x_2]$ onto $[g(x_1), g(x_2)]$ (where $g(x_1) < g(x_2)$) is said to be Cantor's step-function (or devil's staircase) in the interval $[x_1, x_2]$ if for every rational number $\frac{n}{m} \in [g(x_1), g(x_2)]$ and only for a rational one the set

$$I(\tfrac{n}{m}) = \left\{ x : g(x) = \tfrac{n}{m} \right\}$$

is a closed interval (not a point).

Cantor's step-function has the following properties.

(a) If $x_1 < x_2 < x_3 < x_4$ and $g(x)$ is a Cantor's step-function for $x \in [x_1, x_4]$ then it is either a Cantor's step-function or a constant function for $x \in [x_2, x_3]$.

(b) If h is a continuous and increasing mapping of the interval $[y_1, y_2]$ onto $[x_1, x_2]$, and if g is a Cantor's step-function in $[x_1, x_2]$, then $g(h(y))$ is a Cantor's step-function in $[y_1, y_2]$.

(c) Let g_1 and g_2 be two Cantor's step-functions. The first one maps the interval $[x_1, x_2]$ onto $[0, 1]$, and the second maps $[y_1, y_2]$ onto $[0, 1]$. Then there exists such a continuous and increasing mapping h of $[x_1, x_2]$ onto $[y_1, y_2]$ that $g_1(x) = g_2(h(x))$.

It is not evident that Cantor's step-functions exist. Therefore the constructing of examples of such functions can be useful.

2.5.2 *Constructing of a devil's staircase*

We will introduce a set of increasing sequences of rational numbers similar to the Farey tree (compare Ref. [40]).

Lemma 2.1. *If n_1, m_1, n_2, m_2 are natural numbers, and $m_1 n_2 - m_2 n_1 = 1$ (in the point* (b) *the case $n_1 = 0, m_1 = n_2 = 1$ is also acceptable), then:*

(a) *numbers n_1, m_1 are coprime, and n_2, m_2 are coprime.*

(b) *each rational number from the interval $\left(\frac{n_1}{m_1}, \frac{n_2}{m_2} \right)$ can be represented by*

$$\frac{a n_1 + b n_2}{a m_1 + b m_2}, \qquad (2.58)$$

where a, b are natural coprime numbers (numbers $a n_1 + b n_2$ and $a m_1 + b m_2$ are also coprime).

(c) *in the set of numbers* (2.58) *the number*

$$\frac{n_*}{m_*} = \frac{n_1 + n_2}{m_1 + m_2} \qquad (2.59)$$

has the smallest denominator, and the equalities

$$m_1 n_* - m_* n_1 = 1, \qquad m_* n_2 - m_2 n_* = 1 \qquad (2.60)$$

hold.

Proof.   (a) Let there exist a common natural factor w of numbers n_1 and m_1 (or n_2 and m_2). Then the number $m_1 n_2 - m_2 n_1 = 1$ has also the same factor w and, therefore, $w = 1$.

(b) Let n_0, m_0 be coprime natural numbers for which $\frac{n_1}{m_1} < \frac{n_0}{m_0} < \frac{n_2}{m_2}$. If

$$n_0 = a n_1 + b n_2, \qquad m_0 = a m_1 + b m_2,$$

then

$$a = n_2 m_0 - n_0 m_2, \qquad b = n_0 m_1 - n_1 m_0$$

are positive coprime numbers.

(c) For $a = b = 1$ we have

$$m_1(n_1 + n_2) - (m_1 + m_2)n_1 = 1, \qquad (m_1 + m_2)n_2 - m_2(n_1 + n_2) = 1,$$

which gives (2.60). $\square$

Now, let us construct a sequence $E_0, E_1, E_2, E_3 \ldots$ of subsets of rational numbers in the following way:

$$E_0 : \frac{0}{1} \quad \frac{1}{1}$$

$$E_1 : \frac{0}{1} \quad \frac{1}{2} \quad \frac{1}{1}$$

$$E_2 : \frac{0}{1} \quad \frac{1}{3} \quad \frac{1}{2} \quad \frac{2}{3} \quad \frac{1}{1}$$

$$E_3 : \frac{0}{1} \quad \frac{1}{4} \quad \frac{1}{3} \quad \frac{2}{5} \quad \frac{1}{2} \quad \frac{3}{5} \quad \frac{2}{3} \quad \frac{3}{4} \quad \frac{1}{1}$$

$$\ldots \quad . \quad . \quad . \quad . \quad . \quad . \quad . \quad . \quad . \quad . \quad .$$

Every set E_k consists of $(2^k + 1)$ rational numbers from the interval $[0, 1]$ and is constructed as follows. Between any two adjacent numbers $\frac{n_1}{m_1} < \frac{n_2}{m_2}$ of the set E_{k-1} the number $\frac{n_1+n_2}{m_1+m_2}$ is placed. It has the smallest denominator among rational numbers of the interval $(\frac{n_1}{m_1}, \frac{n_2}{m_2})$.

Evidently

$$E_0 \subset E_1 \subset E_2 \subset \ldots \subset E_\infty, \quad \text{where} \quad E_\infty = \bigcup_{0 \le k < \infty} E_k,$$

and E_∞ is the set of all rational numbers of the interval $[0, 1]$. Any two adjacent numbers $\frac{n_1}{m_1} < \frac{n_2}{m_2}$ of the set E_k satisfy the equality $m_1 n_2 - m_2 n_1 = 1$.

Now, let us construct a sequence $I_0, I_1, I_2, I_3 \ldots$ of sets of intervals contained in $[0, 1]$. Let $a \in (0, \frac{1}{3})$ be a fixed number. The first three sets are:

$$I_0 : [0, a], \quad [1 - a, 1]$$

$$I_1 : [0, a], \quad \left[\frac{1-a^2}{2}, \frac{1+a^2}{2} \right], \quad [1 - a, 1]$$

$$I_2 : [0, a], \quad \left[\frac{1+2a-a^2-2a^3}{4}, \frac{1+2a-a^2+2a^3}{4} \right], \quad \left[\frac{1-a^2}{2}, \frac{1+a^2}{2} \right],$$

$$\left[\frac{3-2a+a^2-2a^3}{4}, \frac{3-2a+a^2+2a^3}{4} \right], \quad [1 - a, 1].$$

The set I_k consists of $(2^k + 1)$ closed intervals. It includes all intervals of the set I_{k-1} and additionally 2^{k-1} intervals of the length a^{k+1} which are constructed as follows. Between any two adjacent intervals $[r_1, r_2], [s_1, s_2]$ of the set I_{k-1} the interval $\left[\frac{1}{2}(r_2 + s_1 - a^{k+1}), \frac{1}{2}(r_2 + s_1 + a^{k+1})\right]$ is placed. Evidently,

$$I_0 \subset I_1 \subset I_2 \subset \ldots \subset I_\infty, \quad \text{where } I_\infty = \bigcup_{0 \leq k < \infty} I_k \subset [0,1],$$

and I_∞ is dense in the interval $[0,1]$.

The sum of the lengths of all intervals belonging to I_k is

$$\text{mes } I_k = 2a + a^2 + 2a^3 + \ldots + 2^{k-1}a^{k+1} = 2a + a^2 \frac{1 - 2^k a^k}{1 - 2a} \quad (2.61)$$

and for $k \to \infty$ we have

$$\text{mes } I_\infty = a\frac{2 - 3a}{1 - 2a} < 1 \quad \text{for } a \in \left(0, \frac{1}{3}\right).$$

In every set I_k we define the function g_a putting in successive intervals of I_k the constant values of g_a equal to successive numbers from the set E_k (see Fig.2.14). For $k \to \infty$ we obtain the continuous nondecreasing function g_a which maps the set I_∞ onto the set E_∞ of all rational numbers of the interval $[0,1]$.

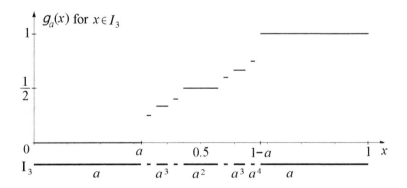

Fig. 2.14   Construction of a devil's staircase.

The domain of g_a can be extended to the whole interval $[0,1]$. The set I_∞ is dense in $[0,1]$. Hence, for every $x_* \in [0,1]/I_\infty$ there exists in I_∞ such a sequence of points $x_1, x_2, x_3, \ldots$ that $\lim_{n \to \infty} x_n = x_*$. Defining

$g_a(x_*) = \lim_{n\to\infty} g_a(x_n)$, we obtain the continuous function which maps the interval $[0,1]$ onto itself.

Additionally, assuming that $g_a(x+1) = g_a(x)+1$, we obtain the function g_a defined for all real numbers.

Remark. In 1883 Georg Cantor showed an example of a continuous nondecreasing function which maps the interval $[0,1]$ onto itself and which almost everywhere has the derivative equal to zero. This example is important for the fundamentals of mathematical analysis. It is surprising that it is also important for the description of some properties of physical systems. A basic characteristic of synchronization has properties similar to those of the Cantor function.

The classical Cantor's example (see Ref. [16]) of the step-function has $a = \frac{1}{3}$ and, consequently, mes $I_\infty = 1$. Moreover, the set of values of this function defined on I_∞ is not the set of all rational numbers from $[0,1]$, but only the set of numbers of the form $m2^{-n}$, where n, m are integers. In construction of the set E_k, between two adjacent numbers $\frac{n_1}{m_1} < \frac{n_2}{m_2}$ of the set E_{k-1} the number $\frac{1}{2}(\frac{n_1}{m_1} + \frac{n_2}{m_2})$ is placed.

The construction of the function g_a given in this section is the same as in Cantor's example with the exception of these two details.

2.5.3 *T-property*

The notion "typical property" is well defined in the theory of dynamical systems (see Ref. [57]), but the definition is not easy. On the other hand in colloquial language the word "typical" is well comprehensible. Now we define one of the properties which is *typical* in strictly mathematical sense and in colloquial language. We say it *T-property*.

Definition 2.6. We say that the continuous function $f(p)$ satisfying condition $f(p+2\pi) = f(p) + 2\pi$ has T-property if there is no natural number m, for which the function $f^m(p) - p$ is a constant function.

It is evident that if a function f_0 has *T-property* and if $\sup_p |f_0(p) - f(p)|$ is sufficiently small, then the function f also has *T-property*.

Example 2.1. The function f which satisfies condition $f(p+2\pi) = f(p) + 2\pi$ and which is defined in the interval $[-\frac{4}{3}\pi, \frac{2}{3}\pi]$ by

$$f(p) = \begin{cases} \frac{3}{2}\pi + \frac{1}{2}p & \text{for} \quad [-\frac{4}{3}\pi, 0] \\ \frac{2}{3}\pi + 2p & \text{for} \quad [0, \frac{2}{3}\pi] \end{cases}$$

does not have *T-property*. It is easily seen (Fig.2.15) that $f^2(p) - p \equiv 2\pi$.

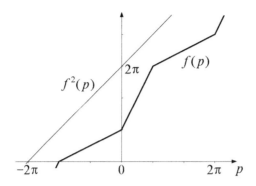

Fig. 2.15   Graph of a function $f(p)$ which does not have T-property.

Example 2.2. If f is not a strictly increasing function then it has *T-property*. In fact, if $f(p_1) = f(p_2)$ for $p_1 \neq p_2$ then for every natural m we have $f^m(p_1) = f^m(p_2)$, and the function $g(p) = f^m(p) - p$ is not a constant function: $g(p_1) \neq g(p_2)$.

Example 2.3. If $F(x) > 0$ for all x, then the solutions of equation (2.18) are given by (2.19)–(2.20). If $s = \frac{n}{m}$ is a rational number, then the Poincaré mapping $f(p)$ does not have *T-property*. In fact, for $p = x(0) = -s\tau_* + \beta(-s\tau_*)$ we have $f^m(p) = x(2\pi m) = 2\pi m s - s\tau_* + \beta(2\pi m s - s\tau_*)$. Then for $s = \frac{n}{m}$ we have $f^m(p) - p = 2\pi n + \big(\beta(2\pi n - s\tau_*) - \beta(-s\tau_*)\big) = 2\pi n$. The last equality is a consequence of periodicity of the function β.

Theorem 2.10. *If*

$$w(p) = A_0 + 2 \sum_{k=1}^{N} A_k \cos(kp + \alpha_k) \tag{2.62}$$

is a trigonometric polynomial not equal to a constant function $(1 \leq N < \infty,$ $A_N \neq 0)$ *then the function* $f(p) = p + w(p)$ *has T-property.*

Proof. The function $f(p) = p + w(p)$, defined for all real values p, can be extended to the whole complex plane such that

$$f(z) = z + \sum_{k=-N}^{+N} c_k e^{ikz}, \quad \text{where} \quad c_k = A_k e^{i\alpha_k} \quad \text{and} \quad c_{-k} = \bar{c}_k,$$

is an entire function (analytic at every point $z \neq \infty$). Consequently, for every natural number m the function $h(z) = f^m(z) - z$ is also an entire function and its derivative is

$$h'(z) = \prod_{k=0}^{m-1} f'\big(f^k(z)\big) - 1.$$

So, if we choose z_* such that $f'(z_*) = 0$ then $h'(z_*) = -1$. Since

$$f'(z) = 1 + \sum_{k=-N}^{+N} ikc_k e^{ikz} = 1 + \sum_{k=-N}^{+N} ikc_k u^k, \quad \text{where} \quad u = e^{iz},$$

we can take $z_* = -i \ln u_*$, where u_* is an arbitrary root of the polynomial

$$A(u) = u^N f'(-i \ln u) = u^N + \sum_{k=-N}^{+N} ikc_k u^{k+N}.$$

Suppose that there exists a natural number m such that $h(p) = f^m(p) - p$ is a constant function for all real values p. Then the entire function $h(z) = f^m(z) - z$ is also the constant function and $h'(z) = 0$ in the whole complex plane, but this contradicts the fact that $h'(z_*) = -1$. So, we conclude that the function $f(p)$ has *T-property*.   □

Taking into account the Weierstrass approximation theorem, we obtain the following important conclusion:

Corollary 2.1. *For every continuous function $f(p)$ satisfying the condition $f(p + 2\pi) = f(p) + 2\pi$, and for every $\varepsilon > 0$, there exists a function $f_\varepsilon(p)$ which has T-property and $\sup_p |f(p) - f_\varepsilon(p)| < \varepsilon$.*

The set of functions $f(p)$ which have T-property is an open and dense set in the space of continuous functions satisfying the condition (2.46).

2.5.4   A fundamental Theorem

Let $\rho(\mu)$ be the rotation number of a continuous mapping $f(p, \mu)$ which satisfies the condition $f(p + 2\pi, \mu) = f(p, \mu) + 2\pi$. Let f be the function nondecreasing with respect to p and increasing with respect to μ for every p and for $\mu \in [\mu_1, \mu_2]$.

Theorem 2.11. *If the function $f(p, \mu)$ has T-property for all $\mu \in [\mu_1, \mu_2]$ and if $\rho(\mu_1) \neq \rho(\mu_2)$, then $\rho(\mu)$ is Cantor's step-function in the interval $[\mu_1, \mu_2]$.*

Proof. For each natural number m the function $f^m(p,\mu) - p$ is continuous. It is periodic with respect to p and increasing with respect to μ. Let us introduce two continuous and increasing functions

$$h_m(\mu) = \min_p \big(f^m(p,\mu) - p\big), \qquad H_m(\mu) = \max_p \big(f^m(p,\mu) - p\big). \qquad (2.63)$$

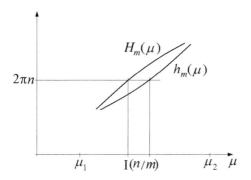

Fig. 2.16   An example of the functions (2.63).

If $f(p,\mu)$ has *T-property* for all $\mu \in [\mu_1, \mu_2]$ then $h_m(\mu) < H_m(\mu)$ (Fig.2.16). For a fixed rational ρ the equality $\rho = \frac{n}{m}$ holds if and only if the number $2\pi n$ belongs to the set of values of the function $f^m(p,\mu) - p$ for $p \in [0, 2\pi]$. Therefore

$$I(n/m) \stackrel{\text{def}}{=} \{\mu : \rho(\mu) = \tfrac{n}{m}\} = \{\mu : h_m(\mu) \le 2\pi n \le H_m(\mu)\}. \qquad (2.64)$$

From inequality $h_m(\mu) < H_m(\mu)$ and from continuity of the function $\rho(\mu)$ we conclude that if $\frac{n}{m} \in \big(\rho(\mu_1), \rho(\mu_2)\big)$ then $I(n/m)$ is a closed interval.

Now we prove that if $\rho(\mu_a)$ is an irrational number then ρ is an increasing function at the point μ_a.

Let $\mu_1 < \mu_a < \mu_b < \mu_2$. As $f(p,\mu)$ is an increasing function of μ, then there exists $\varepsilon > 0$ such that for every p we have $f(p,\mu_a) + \varepsilon < f(p,\mu_b)$. Consequently, $f^m(p,\mu_a) + \varepsilon < f^m(p,\mu_b)$ for every natural number m. Let us assume (for simplicity of the proof) that $f(p,\mu_a)$ has a continuous and positive derivative with bounded variation. Then, by Theorem 2.6, the trajectory $\{f^m(0,\mu_a)\}$ with the irrational rotation number $\rho(\mu_a)$ is dense in the circle. For arbitrary m_0 there exist two natural numbers m, n such that $m > m_0$ and $-\varepsilon < f^m(0,\mu_a) - 2\pi n < 0$. This gives

$$\frac{f^m(0,\mu_a)}{2\pi m} < \frac{n}{m} \quad \text{and} \quad \frac{f^m(0,\mu_b)}{2\pi m} > \frac{n}{m},$$

and, consequently, $\rho(\mu_a) < \frac{n}{m} \leq \rho(\mu_b)$. The left-hand side inequality (less then) is a consequence of the fact that $\rho(\mu_a)$ is an irrational number. So, for every irrational number l the set $\{\mu : \rho(\mu) = l\}$ consists of one point only.

This finishes the proof. $\square$

2.5.5   *Consequences for forced oscillators*

Let $U_{\text{out}}\big(\theta(t)\big)$ be the output signal of an oscillator controlled by a periodic signal $U_{\text{inp}}(\omega t)$. Various types of oscillators can be described by the differential equation

$$\frac{d\theta}{d\tau} = F(\theta, \tau; \mu, A), \tag{2.65}$$

where F is a positive 2π-periodic function of both variables θ, τ and it depends on two positive parameters μ, A. These variables and parameters have the following physical meaning:

$\tau = \omega t$ is a phase, and ω is a frequency of the periodic input signal $U_{\text{inp}}(\tau) = U_{\text{inp}}(\tau + 2\pi)$,

A is an amplitude (the peak value) of the input signal,

θ is a phase of the periodic output signal $U_{\text{out}}(\theta) = U_{\text{out}}(\theta + 2\pi)$, and if $A = 0$ then $\theta = \Omega t$, where Ω is a self-frequency of the autonomous oscillator,

$\mu = \Omega/\omega$ is the quotient of two frequencies.

A concrete function F depends on properties of oscillators and on the periodic functions U_{inp}, U_{out}, but for $A = 0$ we have $F(\theta, \tau; \mu, 0) \equiv \mu$.

A simple example of such an oscillating system was given in Section 2.1, but equations of numerous more complicated oscillators can be also reduced to equation (2.65).

The output signal can be characterized by the instantaneous frequency

$$\frac{d\theta}{dt} = \omega \frac{d\theta(\tau)}{d\tau},$$

which is a function of time, and by the center-frequency ω_c which is defined as the mean-value of the instantaneous frequency

$$\omega_c = \lim_{\tau \to \infty} \frac{1}{\tau} \int_0^\tau \omega \frac{d\theta(\tau')}{d\tau'} d\tau' = \omega \lim_{\tau \to \infty} \frac{\theta(\tau) - \theta(0)}{\tau} = \lim_{t \to \infty} \frac{\theta}{t}. \tag{2.66}$$

The ratio ω_c/ω is equal to the rotation number of equation (2.65)

$$\rho(\mu, A) = \frac{\text{center-frequency of output signal}}{\text{input frequency}}, \tag{2.67}$$

and it depends on parameters μ, A. The function (2.67) is the basic characteristic of synchronized oscillators. We are especially interested in the so called *Arnold's tongues* (see Ref. [40]).

Definition 2.7. Let $\frac{n}{m}$ be a given rational number. In the plane of parameters μ, A the set

$$Q_{n/m} = \left\{ (\mu, A) \ : \ \rho(\mu, A) = \frac{n}{m} \right\} \qquad (2.68)$$

is called the Arnold tongue of the number $\frac{n}{m}$.

For every $(\mu, A) \in Q_{n/m}$ there exists a periodic point p_0 such that the graph of solution $\theta(\tau; p_0)$ of equation (2.65) with the initial value $\theta(0, p_0) = = p_0$ is a closed trajectory on the torus. This trajectory (periodic orbit) surrounds the torus m-times in direction of τ-axis, and n-times in direction of θ-axis. The output signal $U_{\text{out}}\big(\theta(\tau; p_0)\big)$ is $2\pi m$-periodic with respect to τ and $\theta(2\pi m; p_0) - \theta(0; p_0) = 2\pi n$.

This output signal is not necessarily stable. For example, if (μ, A) belongs to the boundary of $Q_{n/m}$ then p_0 is a neutral periodic point of the type n/m. However, if (μ, A) belongs to the interior of Arnold's tongue then there exists at least one pair of periodic orbits: stable and unstable. The existence of a stable periodic orbit of the type n/m (where $m \geq 2$) can be interpreted as a *fractional synchronization*. So, the interior of Arnold's tongue is also called the *region of synchronization*.

2.5.6 Numerical and analytical approach

The following numerical results illustrate the properties given above for the equation (2.3).

An approximate formula for the Poincaré mapping $f(p; \mu, A)$ is calculated with an error of the order $O(A^2)$ (see Section 2.7.1). Using this formula, the graph of the rotation number $\rho(\mu, A)$ is numerically calculated and shown in Fig.2.17a versus $\mu = \frac{\Omega}{\omega}$ for a fixed value $A = 0.1$.

For a few ratios $\frac{n}{m}$ we observe the intervals $I(n/m) = \{\mu : \rho = \frac{n}{m}\}$. If μ belongs to the interior of $I(n/m)$ then there exists a stable periodic orbit of the type n/m.

The graph of a so called *steady-state trajectory* of the Poincaré mapping is shown in Fig.2.17b . It is the graph of the set of points

$$\theta(2\pi k; p, \mu, A) \ mod \ 2\pi \quad \text{for} \quad k_{\min} \leq k \leq k_{\max} \qquad (2.69)$$

versus μ, where p is a random initial value and integers $k_{\min}$ and $k_{\max} - k_{\min}$ are sufficiently large. If the system has a stable periodic orbit of the type

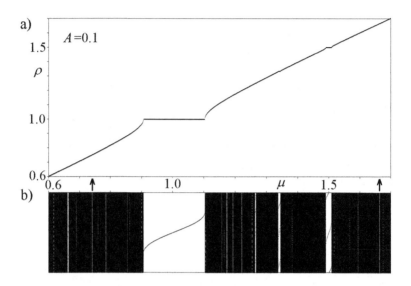

Fig. 2.17   Center-frequency versus self-frequency for the oscillator described by Eq. (2.3) and the steady-state trajectory of the Poincaré mapping.

n/m then we observe that the steady-state trajectory has exactly m different points, provided that $k_{\min} \to \infty$ and $(k_{\max} - k_{\min}) \to \infty$.

Two values of parameter μ are marked in the figure. For these values, two locally stable output signals were calculated using a numerical method of moderate precision. In Fig.2.18 the graphs of input and output signals are shown for the time intervals equal to one period of the output signal. The amplitude-spectra of output signals

$$\cos\theta(\tau) = \sum_{k=-\infty}^{+\infty} c(\nu_k)e^{2\pi j\nu_k \tau}, \quad \text{where } \nu_k = \frac{k}{m}, \quad \tau = \omega t$$

are also shown in the figure.

However, if we use more accurate numerical methods for solving the equation (2.3), then not all intervals $I(n/m)$ shown in Fig.2.17 are observed. Probably, the Poincaré mapping of equation (2.3) has no *T-property*. It is not possible to check it by numerical methods because in every neighborhood of an arbitrary Poincaré mapping there exist functions which have *T-property* (see Theorem 2.10). Unfortunately, we do not know any concrete example of equation (2.65), with an elementary function F, for which we can prove that its Poincaré mapping has *T-property*.

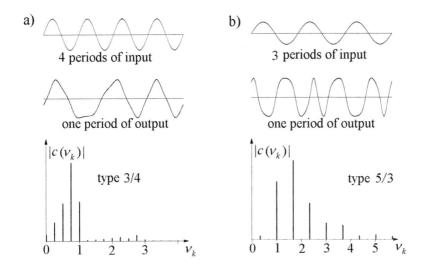

Fig. 2.18   Examples of two stable output signals for the oscillator described by Eq. (2.3) and their amplitude-spectra.

Now we give the outline of an analytical algorithm to determine the set of parameters μ, A for which the rotation number is equal to a given number ρ. Let us introduce new variables ψ, τ such that the map $(\theta, \tau) \to (\psi, \tau)$ is a homeomorphism of the torus $S \times S$ onto itself and for the new variables the equation (2.3) takes the form $\frac{d\psi}{d\tau} = \rho$. This is possible for irrational numbers ρ (see Theorem 2.11), and maybe also for some rational numbers, provided the Poincaré mapping has no *T-property*. It is convenient to seek the function $\psi(\theta, \tau)$ as the Fourier expansion

$$\psi = \theta + \sum_{n,m} h_{n,m} e^{j(m\tau + n\theta)}, \quad \tau = \omega t. \tag{2.70}$$

Substituting (2.70) to the equation (2.3) and assuming $\frac{d\psi}{d\tau} = \rho$, we obtain an infinite number of equations for all coefficients $h_{n,m}$. For a fixed ρ, these equations depend on the parameters A and $\mu = \frac{\Omega}{\omega}$, and some relationships between μ and A are necessary for the existence of solutions.

For small values of A it is convenient to expand the coefficients $h_{n,m}(\rho, A)$ and $\mu(\rho, A)$ into the formal series with respect to the powers of A. Next, we recurrently calculate the coefficients $R_i(\rho)$ of the formal series

$$\mu(\rho, A) = \rho + R_1(\rho)A + R_2(\rho)A^2 + \ldots + R_n(\rho)A^n + \ldots \tag{2.71}$$

for which there exist the Fourier coefficients $h_{n,m}$. The functions $R_i(\rho)$ take the form of rational functions. If at least one function $R_i(\rho)$ has a pole at a point ρ_0, then we can not reduce the equation (2.3) to $\frac{d\psi}{d\tau} = \rho_0$, and probably there exists the Arnold tongue of the number ρ_0.

Restricting the calculation to accuracy $O(A^5)$ we obtain the formula

$$\mu(\rho, A) = \rho + \frac{\rho^3}{\rho^2 - 1} A^2 + \frac{\rho^5(7\rho^2 - 11)}{4(\rho^2 - 1)^3} A^4 + O(A^6) \quad \text{for } \rho \neq 1, 3 \quad (2.72)$$

as a necessary condition of existence of such Fourier's coefficients $h_{n,m}$ for which the equation (2.3) takes the form

$$\frac{d\psi}{d\tau} = \rho + O(A^5), \qquad \rho \neq 1, 3. \qquad (2.73)$$

For $\rho = 1, 3$ (and probably for $\rho = 5, 7, 9, ...$) the equation (2.3) can not be reduced to the form $\frac{d\psi}{d\tau} = \rho$, but it can be reduced to the equation $\frac{d\psi}{d\tau} = \rho + F(\psi - \rho\tau)$ where F is a 2π-periodic function which depends on μ and A. Namely

$$\frac{d\psi}{d\tau} = 1 + \left(\mu - 1 - \tfrac{1}{4}A^2\right) - \mu A \sin(\psi - \tau) + O\left(A^3\right) \qquad \text{for } \rho = 1,$$

$$\frac{d\psi}{d\tau} = 3 + \left(\mu - 3 - \tfrac{27}{8}A^2\right) + \tfrac{27}{16}A^3 \sin(\psi - 3\tau) + O\left(A^4\right) \quad \text{for } \rho = 3. \qquad (2.74)$$

These equations can be reduced to the autonomous ones $\frac{dx}{d\tau} = F(x)$ by substitution $\psi = x + \rho\tau$, and the results of the Section 2.2 can be used.

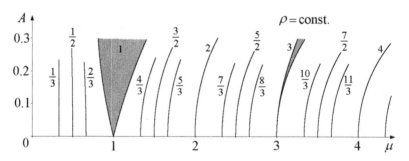

Fig. 2.19   The sets of parameters μ, A, for which the rotation number ρ of Eq. (2.3) takes a constant value.

From the above it follows that for $\rho = 1$ and for $\rho = 3$ there exist the Arnold tongues:

$$\rho = 1 \quad \text{for} \quad |\mu - 1 - \tfrac{5}{4}A^2| \leq A + O(A^3),$$

$$\rho = 3 \quad \text{for} \quad |\mu - 3 - \tfrac{27}{8}A^2| \leq \tfrac{27}{16}A^3 + O(A^4). \qquad (2.75)$$

Since we calculate with a finite accuracy $O(A^5)$, we do not know whether the Arnold tongues exist for other values of ρ or not, but if they exist then they are very narrow: if the rotation number ρ takes a constant value for $\mu \in [\mu_1(A), \mu_2(A)]$ then we have $|\mu_2(A) - \mu_1(A)| \leq O(A^5)$ for $\rho \neq 1, 3$.
The results given by (2.72) and (2.75) are shown in Fig.2.19.

2.6 The system with rectangular waveform signals

Let the input and output signals of PLL system be *rectangular waveform*:

$$U_{\text{inp}}(\omega t) = 2AV R(\omega t), \quad U_{\text{out}}\big(\theta(t)\big) = R\big(\theta(t)\big), \quad (2.76)$$

where $R(\tau) = \text{sgn}(\sin \tau)$.
In this case the system is described by the differential equation

$$\frac{d\theta}{d\tau} = \mu\big(1 + 2AR(\tau) R(\theta)\big), \quad \tau = \omega t \quad (2.77)$$

which depends on two parameters $\mu = \frac{\Omega}{\omega}$ and $2A \in (0, 1)$.
The solution $\theta(\tau; \tau_0, \theta_0)$ with an initial value (τ_0, θ_0) is a piecewise linear function with slope

$$\frac{d\theta}{d\tau} = \begin{cases} C \text{ for } R(\tau) R(\theta) < 0 \\ D \text{ for } R(\tau) R(\theta) > 0 \end{cases} \quad (2.78)$$

where

$$C = \mu(1 - 2A), \quad D = \mu(1 + 2A). \quad (2.79)$$

Therefore we can get an effective formula for the Poincaré mapping and present a geometrical algorithm for calculating the Arnold tongues.

2.6.1 *The Poincaré mapping*

Using the symmetry $R(\tau) = -R(\tau + \pi)$ of the function R it is convenient to define the Poincaré mapping for equation (2.77) (expressed in the new coordinates $\theta + \tau$, $\theta - \tau$) as a map which transfers points (τ_0, θ_0) of the line $L_0 = \{(\tau, \theta) : \tau + \theta = \pi\}$ along the solutions of equation (2.77) to points (τ_1, θ_1) of the line $L_1 = \{(\tau, \theta) : \tau + \theta = 3\pi\}$.
Let us introduce the coordinate $u = \frac{1}{2\pi}(\theta - \tau + \pi)$ on the lines L_0, L_1. Using the geometrical properties of graphs of solutions shown in Fig.2.20a,

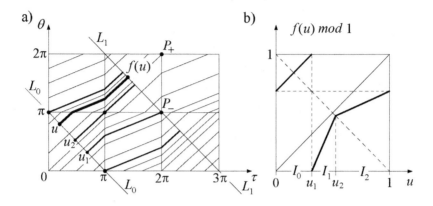

Fig. 2.20   Graph of a family of solutions of Eq. (2.77), and the Poincaré mapping (2.80).

it is easily seen that for $\mu < 1$ the Poincaré mapping is of the form

$$f(u) = \begin{cases} u + \frac{C+D}{1+D} & \text{for} \quad u \in I_0 = \left[0, \frac{1-C}{1+D}\right], \\[2mm] \frac{D}{C}\left(u - \frac{1-C}{1+D}\right) + 1 & \text{for} \quad u \in I_1 = \left[\frac{1-C}{1+D}, \frac{1}{1+D}\right], \\[2mm] \frac{C}{D}(u-1) + \frac{C+D}{1+D} + 1 & \text{for} \quad u \in I_2 = \left[\frac{1}{1+D}, 1\right]. \end{cases} \quad (2.80)$$

The function f is a piecewise linear map contracting for $u \in I_2$, stretching for $u \in I_1$ and translating for $u \in I_0$ (see Fig.2.20b). It fulfils the relation

$$f(u+1) = f(u) + 1 \quad \text{for} \quad -\infty < u < +\infty.$$

Moreover, from the symmetry of the function R it follows that

$$f(u) + f^{-1}(1-u) = 1 \, mod \, 1. \quad (2.81)$$

The change of coordinates $(\tau, \theta) \to (\theta + \tau, \theta - \tau)$ in equation (2.77) leads to the relation

$$\rho = \frac{1 + \rho_0}{1 - \rho_0}$$

between the rotation number ρ of equation (2.77) and the rotation number ρ_0 of the Poincaré mapping f. Indeed

$$\rho_0 = \lim_{\tau \to \infty} \frac{\theta - \tau}{\theta + \tau} = \lim_{\tau \to \infty} \frac{\frac{\theta}{\tau} - 1}{\frac{\theta}{\tau} + 1} = \frac{\rho - 1}{\rho + 1},$$

and if $0 < \rho < 1$ then $-1 < \rho_0 < 0$.

If $-\rho_0 = \frac{n}{m}$ (where n, m are coprime numbers) then the map $f(u) \, mod \, 1$ has a periodic trajectory consisting of m points, where n of them belong to the interval I_0.

If $m - n = 2k + 1 \geq 1$ is an odd number then there exist two periodic trajectories of the type n/m: a stable and an unstable one. The points of the stable and unstable periodic trajectories occur alternately in the interval $u \in [0, 1]$. From the symmetry (2.81) it follows that the unstable trajectory has $k + 1$ points in the interval I_1 and k points in the interval I_2, while the stable trajectory has k points in the interval I_1 and $k + 1$ points in the interval I_2. The rotation number of equation (2.77)

$$\rho = \frac{1 + \rho_0}{1 - \rho_0} = \frac{1 - \frac{n}{m}}{1 + \frac{n}{m}} = \frac{m - n}{m + n} = \frac{2k + 1}{2(k + n) - 1} \quad (2.82)$$

is then the number such that its nominator and denominator are both odd coprime numbers.

In the plane of parameters μ, A there exists the Arnold's tongue Q_ρ i.e. the closure of domain of points (μ, A) for which the equation (2.77) has the rotation number (2.82) and a stable periodic orbit. On the boundary of domain Q_ρ there occurs the bifurcation of joining of the stable and unstable periodic trajectories. The equations of this boundary can be obtained from the following conditions: there exists the periodic trajectory of the map $f(u) \, mod \, 1$ (Fig.2.20b) containing the point $u_2 = \frac{1}{1+D}$ for the curve bounding the Arnold's tongue from the left, and the point $u_1 = \frac{1-C}{1+D}$ for the curve bounding the Arnold's tongue from the right. It is evident that the point u_1 and the point $u = 1$ correspond the same trajectory on the torus.

If $m - n = 2k > 0$ is an even number or, equivalently, if m and n are coprime odd numbers, then the map $f(u) \, mod \, 1$ (Fig.2.20b) has a periodic trajectory consisting of m points, where n of them belong to the interval I_0, k belong to the interval I_1 and k to the interval I_2. This periodic trajectory is neutral (provided that it does not contain points u_1 and u_2). From the piecewise linearity of the map (2.80) and its symmetry (2.81) it follows that the periodic trajectory shifted by any interval on the circle is also a periodic neutral trajectory. It also means that the m-th iteration $f^m(u)$ of the Poincaré mapping is a translation by the interval equal to $-n$. The rotation number of the equation (2.77)

$$\rho = \frac{1 + \rho_0}{1 - \rho_0} = \frac{1 - \frac{n}{m}}{1 + \frac{n}{m}} = \frac{m - n}{m + n} = \frac{k}{k + n} \quad (2.83)$$

is an irreducible ratio where the nominator or denominator is an even number (because n is an odd number).

In such case the Arnold's tongue reduces to the curve in μA-plane. The equation of this curve can be obtained from the following condition: there exists the periodic trajectory of the map $f(u) \bmod 1$ which contains simultaneously the points u_1 and u_2.

2.6.2   The Arnold's tongues

In $\tau\theta$-plane the points with coordinates $(p\pi, q\pi)$ (where p, q integers) are of two types: of the type P_+ if $(p+q)$ is an even number, and of the type P_- if $(p+q)$ is an odd number (Fig.2.20a). The periodic trajectory of the map $f(u) \bmod 1$ contains the point u_2 (or u_1) if and only if the corresponding periodic solution of equation (2.77) passes in $\tau\theta$-plane through the points of type P_+ (or P_- respectively). It follows that a point (μ, A) belongs to the left-hand side (or right-hand side) boundary of the Arnold's tongue $Q_{p/q}$ if and only if there exists a periodic solution of the equation (2.77) passing through the points of type P_+ (or P_- respectively) and if the straight line connecting these points has the slope $\frac{p}{q}$. If p or q is an even number then the above mentioned periodic solution passes through both points P_+ and P_-, and the Arnold's tongue reduces to the line.

Example 2.4. In μA-plane we determine the line where the equation (2.77) has the rotation number $\rho = \frac{2}{3}$.

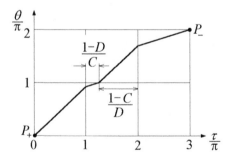

Fig. 2.21   Periodic solution of Eq. (2.77) of the type $2/3$ passing through the points P_+ and P_- (for $A = 0.25$).

In Fig.2.21 the periodic solution of (2.77) is shown with the given rotation number and passing through the points P_+ and P_-. The geometrical construction implies the following condition of the existence of this solution:

$$\frac{1-D}{C} + \frac{1-C}{D} = 1.$$

Using (2.79) we conclude that $\rho = \frac{2}{3}$ is the rotation number of the equation (2.77) for

$$\mu = \frac{2}{3 + 4A^2}.$$

It is a line in μA-plane.

Example 2.5. Let us determine in μA-plane the equations of lines which bound the Arnold's tongue of the number $\rho = \frac{3}{5}$.

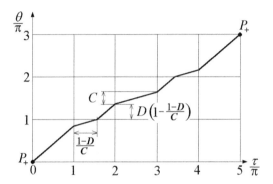

Fig. 2.22   Periodic solution of Eq. (2.77) of the type $3/5$ passing through the points P_+ (for $A = 0.25$).

In Fig.2.22 the periodic solution of (2.77) is shown with the given rotation number and passing through the points P_+. Such a solution occurs for the parameters μ, A lying on the left-hand side boundary of Arnold's tongue. The geometrical condition of existence of this solution is

$$2D \left(1 - \frac{1-D}{C} \right) + C = 1.$$

Interchanging C and D in the above equation we obtain the condition of the existence of the periodic solution which passes through the points P_- and which occurs for the parameters μ, A lying on the right-hand side boundary of Arnold's tongue. Using (2.79) we conclude that for

$$\frac{3 + 2A}{5 + 4A + 4A^2} \leq \mu \leq \frac{3 - 2A}{5 - 4A + 4A^2}$$

the equation (2.77) has the Arnold's tongue of the number $\rho = \frac{3}{5}$.

The geometrical method illustrated by the above examples can be used for determination of numerous Arnold's tongues. In particular

$$\rho = \frac{1}{2k+1} \quad \text{for} \quad \frac{1}{2k+1+2A} \leq \mu \leq \frac{1}{2k+1-2A} \tag{2.84}$$

and $\rho = \frac{1}{2k}$ for $\mu = \frac{1}{2k}$, where $k = 1, 2, 3, \ldots$

2.6.3   Numerical results and consequences of a symmetry

Numerical methods can be also used for calculating the rotation number

$$\rho(\mu, A) = \lim_{\tau \to \infty} \frac{\theta(\tau; p, \mu, A) - p}{\tau} \tag{2.85}$$

of the equation (2.77). In Fig.2.23 there are shown the Arnold's tongues of a few numbers $\frac{n}{m}$ (where n and m are odd numbers) in the domain $\mu \in (0.08, 1)$, $A \in (0, 0.5)$.

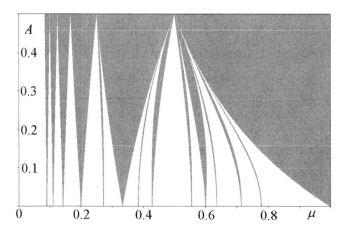

Fig. 2.23   Arnold's tongues of the numbers 1/11, 1/9, 1/7, 1/5, 3/11, 1/3, 5/13, 3/7, 5/9, 3/5, 7/11, 5/7, 7/9, 1/1 for Eq. (2.77).

For $\mu > 1$ the Arnold's tongues can be determined using a special symmetry of the rotation number of equation (2.77). Let $\rho_1(C, D)$ be the rotation number of the equation

$$\frac{dy}{dx} = \begin{cases} C & \text{for } R(x)\, R(y) < 0 \\ D & \text{for } R(x)\, R(y) > 0 \end{cases} \quad \text{where } C > 0,\ D > 0. \tag{2.86}$$

It is evident that

$$\rho_1(C,D) = \lim_{x\to\infty} \frac{y}{x} = \left(\lim_{y\to\infty} \frac{x}{y}\right)^{-1} = \frac{1}{\rho_1\left(C^{-1}, D^{-1}\right)} \qquad (2.87)$$

and that $\rho_1(C,D) = \rho_1(D,C)$.

Between the rotation numbers of equations (2.77) and (2.86) we have the following relations:

$$\rho(\mu, A) = \rho_1(C,D),$$

where

$$C = \mu(1-2A), \quad D = \mu(1+2A), \quad \text{or} \quad \mu = \frac{C+D}{2}, \quad A = \frac{|C-D|}{2(C+D)}. \qquad (2.88)$$

If

$$\rho(\mu_1, A_1) = \frac{n}{m} \quad \text{for} \quad \mu_1 = \frac{C+D}{2} \quad \text{and} \quad A_1 = \frac{|C-D|}{2(C+D)}$$

then, by (2.87), we have

$$\rho(\mu_2, A_2) = \frac{m}{n} \quad \text{for} \quad \mu_2 = \frac{C^{-1}+D^{-1}}{2} \quad \text{and} \quad A_2 = \frac{|C^{-1}-D^{-1}|}{2(C^{-1}+D^{-1})}.$$

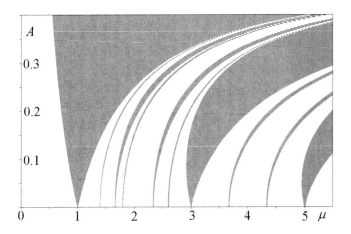

Fig. 2.24 Arnold's tongues of the numbers 1/1, 7/5, 5/3, 9/5, 7/3, 13/5, 3/1, 11/3, 13/3, 5/1 for Eq. (2.77).

Therefore,

$$A_2 = A_1, \quad \text{and} \quad \mu_2 = \frac{1}{\mu_1(1 - 4A_1^2)}.$$

For a fixed A we have the following relation between the two steps:

$$\text{if } I(n/m) = [a, b], \quad \text{then } I(m/n) = \left[\frac{1}{b(1 - 4A^2)}, \frac{1}{a(1 - 4A^2)} \right]. \quad (2.89)$$

Using the formula (2.89) and the results shown in Fig.2.23 we obtain Arnold's tongues for $\mu > 1$ as shown in Fig.2.24.

2.7   The mapping $f(p) = p + 2\pi\mu + a \sin p$

The Poincaré mapping for the equation (2.65) is of the form $f(p) = p + w(p)$, where w is a 2π-periodic function and $f'(p) > 0$. The function $w(p)$ can be effectively calculated by numerical methods only (apart from exceptional cases), and usually it is approximated by a trigonometric polynomial. So it is useful to present some qualitative properties of Arnold's tongues and stable periodic orbits in the case where $w(p)$ is a trigonometric polynomial.

The simplest periodic function contains only the first harmonic and a constant component. Therefore, the family of mappings

$$f(p; \mu, a) = p + 2\pi\mu + a \sin p \quad \text{for } |a| < 1, \quad (2.90)$$

which depend on two parameters μ, a, is worth to be investigated (see Refs. [6], [4]). An additional argument is given below.

2.7.1   *Small input signal*

Let us consider the differential equation

$$\frac{d\theta}{d\tau} = \mu(1 + 2A \sin \tau \cos \theta), \quad \theta(0) = p' \quad (2.91)$$

with a small parameter A. We can seek the solution as a formal power series

$$\theta(\tau; p') = \theta_0(\tau) + A\theta_1(\tau) + A^2\theta_2(\tau) + \dots \quad (2.92)$$

Substituting (2.92) for θ in (2.91) and comparing terms with the same power of A we obtain the set of equations

$$\frac{d\theta_0}{d\tau} = \mu, \quad \frac{d\theta_1}{d\tau} = 2\mu \sin \tau \cos \theta_0(\tau), \quad \frac{d\theta_2}{d\tau} = -2\mu\theta_1(\tau) \sin \tau \sin \theta_0(\tau) \ \dots$$

for all coefficients $\theta_k(\tau)$ of the series. Let us establish the initial conditions: $\theta_0(0) = p'$ and $\theta_k(0) = 0$ for $k = 1, 2, 3, \dots$

A simple computation gives the Poincaré mapping

$$\theta(2\pi; p') = p' + 2\pi\mu + A\frac{4\mu \sin \pi\mu}{1 - \mu^2} \sin(p' + \pi\mu) + O(A^2). \qquad (2.93)$$

If we neglect the term $O(A^2)$ and denote

$$p = p' + \pi\mu, \quad \text{and} \quad f(p; \mu, a) = \theta(2\pi; p') + \pi\mu,$$

we obtain the mapping (2.90) with the parameter $a = 4A\mu(1 - \mu^2)^{-1} \sin \pi\mu$ for $\mu \neq 1$ and $a = 2\pi A$ for $\mu = 1$.

The numerical results shown in Fig.2.17 were obtained using the approximate expressions (2.93) of the Poincaré mapping (for $A = 0.1$).

2.7.2 Properties of the rotation number

Now we give some basic properties of the rotation number

$$\rho(\mu, a) = \lim_{m \to \infty} \frac{f^m(p; \mu, a)}{2\pi m} \qquad (2.94)$$

of the mapping (2.90).

Theorem 2.12. *If $|a| \leq 1$, then the function (2.94) has the following properties:*

(a) *for fixed $a \neq 0$ the function $\rho(\mu, a)$ is Cantor's step-function,*
(b) $\rho(\mu, -a) = \rho(\mu, a)$,
(c) $\rho(\mu + 1, a) = \rho(\mu, a) + 1$,
(d) $\rho(\mu, a) + \rho(1 - \mu, a) = 1$,
(e) *if $|a| \geq 2\pi|\mu - n|$ then $\rho(\mu, a) = n$,*
(f) $\rho(n + \frac{1}{2}, a) = n + \frac{1}{2}$,
(g) $\rho(\mu, 0) = \mu$.

Proof. (a) is a consequence of theorems 2.10 and 2.11.

(b) From $f(p; \mu, -a) = f(p + \pi; \mu, a) - \pi$ we deduce (by induction) that $f^m(p; \mu, -a) = f^m(p + \pi; \mu, a) - \pi$ and from the definition (2.94) it follows that $\rho(\mu, -a) = \rho(\mu, a)$.

(c) From $f(p; \mu + 1, a) = f(p; \mu, a) + 2\pi$ we can prove by induction that $f^m(p; \mu + 1, a) = f^m(p; \mu, a) + 2\pi m$ and from the definition (2.94) it follows that $\rho(\mu + 1, a) = \rho(\mu, a) + 1$.

(d) From $f(p; -\mu, a) = -f(-p; \mu, a)$ it follows that $f^m(p; -\mu, a) = -f^m(-p; \mu, a)$ and, consequently, $\rho(-\mu, a) = -\rho(\mu, a)$. From (c) we have $\rho(-\mu, a) = \rho(1 - \mu, a) - 1$. Combining these two equalities we obtain the property (d).

(e) The equality $\rho(\mu, a) = n$ holds if and only if there exists a solution of equation $f(p; \mu, a) = p + 2\pi n$. Using (2.90) we obtain the property (e).

(f) follows immediately from (c) and (d).

(g) For $a=0$ we have $f^m(p;\mu,0)=p+2\pi m\mu$ and, consequently, $\rho(\mu,0) = = \mu$.

This finishes the proof.   □

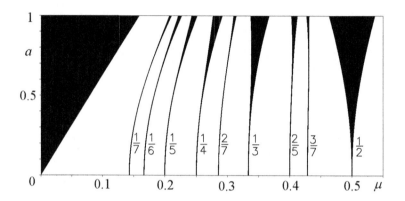

Fig. 2.25   Arnold's tongues of the numbers n/m, where $m \leq 7$ for the mapping (2.90).

In Fig. 2.25 there are shown Arnold's tongues of the numbers $\frac{n}{m} = = \frac{0}{1}, \frac{1}{7}, \frac{1}{6}, \frac{1}{5}, \frac{1}{4}, \frac{2}{7}, \frac{1}{3}, \frac{2}{5}, \frac{3}{7}, \frac{1}{2}$ for the mapping (2.90).

According to the next theorem the Arnold's tongues are very narrow for small values of $|a|$ and for large numbers m (see Ref. [5]).

Theorem 2.13. *Let the mapping* $f(p) = p + 2\pi\mu + \varepsilon w(p)$ *be given, where* $w(p)$ *is a trigonometric polynomial* (2.62) *of degree* N *and* $f'(p) > 0$.

If for $\mu \in [\mu_1(\varepsilon), \mu_2(\varepsilon)]$ *the rotation number* (2.94) *takes a constant value* $\rho(\mu) = \frac{n}{m}$ *then*

$$|\mu_2(\varepsilon) - \mu_1(\varepsilon)| \leq C\varepsilon^r \quad \text{with such integer } r \text{ that } \frac{m}{N} \leq r < \frac{m}{N} + 1, \quad (2.95)$$

where C *does not depend on* ε.

The proof is omitted.

Let us consider the mapping (2.90) for the parameters μ, a belonging to the rectangular $R_\varepsilon = \{(\mu, a) : 0 < \mu < 1, 0 < a < \varepsilon\}$. For each rational number $\frac{n}{m}$ the mapping (2.90) has Arnold's tongue $Q_{n/m}$ with non empty interior. Its measure can be estimated by

$$\text{mes}\left(Q_{n/m} \cap R_\varepsilon\right) \leq \int_0^\varepsilon Ca^m da = \frac{C}{m+1}\varepsilon^{m+1}.$$

In the rectangular R_ε the measure of all Arnold's tongues $Q_{n/m}$ for $m \geq 2$ is estimated by

$$\sum_{m=2}^{\infty} E(m) \frac{C}{m+1} \varepsilon^{m+1} < C \sum_{m=2}^{\infty} \frac{m-1}{m+1} \varepsilon^{m+1} < \frac{C\varepsilon^3}{1-\varepsilon} = O\left(\varepsilon^3\right), \quad (2.96)$$

where $E(m)$ is the Euler function i.e. the number of fractions $\frac{n}{m}$, where n,m are coprime numbers and $1 \leq n \leq m - 1$.

Let S_ε denote the set of such points of the rectangular R_ε which belong to regions of fractional synchronization (to interiors of Arnold's tongues $Q_{n/m}$ for $m \geq 2$). For small input signals the fractional synchronization holds with very small probability:

$$\frac{\mathrm{mes}\, S_\varepsilon}{\mathrm{mes}\, R_\varepsilon} \leq O\left(\varepsilon^2\right). \quad (2.97)$$

Similar results can be given also for differential equations (see Ref. [5]).

Theorem 2.14. *Let $F(x,\tau)$ be a trigonometric polynomial of degree N with respect to x. Let $\rho(\mu)$ denote the rotation number of the differential equation*

$$\frac{dx}{d\tau} = \mu + \varepsilon F(x,\tau), \quad \text{where} \quad F(x+2\pi,\tau) = F(x,\tau) = F(x,\tau+2\pi). \quad (2.98)$$

If for $\mu \in [\mu_1(\varepsilon), \mu_2(\varepsilon)]$ the rotation number takes a constant value $\rho(\mu) = \frac{n}{m}$ then the estimation (2.95) holds.

The proof is omitted.

Consequently, the inequality (2.97) holds also for the differential equation (2.91) with a small value $\varepsilon = A$.

2.7.3   The number of periodic orbits

In the paper [26] M.V. Jakobson proves the following theorem.

Theorem 2.15. *Let $w(p)$ be a trigonometric polynomial (2.62) of degree N. The mapping $f(p) = p + w(p)$, where $f'(p) > 0$ has no more then $2N$ periodic orbits.*

The proof is omitted. It uses the theory of the functions of complex variables and Montel's Theorem on the normal family of meromorphic functions (see also Refs. [3], [11]).

If the mapping $f(p)$ (where $f'(p) > 0$) has no neutral orbits then the number of stable orbits is the same as the number of unstable orbits. So, the mapping (2.90) has no more then one stable and one unstable periodic orbit. Moreover, the proof of Jakobson's theorem leads to some additional properties of the mappings determined by a trigonometric polynomial.

Proposition 2.4. *Let* $w(p)$ *be a trigonometric polynomial* (2.62) *of degree* N. *Let* k *be the number of points* $p_1, ..., p_k$ *of the interval* $[0, 2\pi)$ *for which* $f'(p_i) = 0$. *Assume that* $f''(p_i) \neq 0$ *for each* p_i, $i = 1, ..., k$ *and, consequently,* k *is an even number. Then the mapping* $f(p) = p + w(p)$ *has no more then* $N + k/2$ *stable periodic orbits.*

Properties of the mapping $f(p) = p + 2\pi\mu + a \sin p$, where $|a| > 1$, will be discussed in Chapter 4, but from the above proposition it follows immediately that f has no more then two stable periodic orbits. Consequently, the PLL system described by the Poincaré mapping (2.90) has no more then one stable periodic output signal $U_{\text{out}}(\theta(t))$ for $0 < |a| < 1$ and no more then two such signals for $|a| > 1$.

Chapter 3

The second order continuous-time Phase-Locked Loops

3.1 The system with a low-pass filter

In this chapter we analyze the phase-locked loop (PLL) which contains three units: the voltage controlled oscillator (VCO), the low-pass filter with the transfer function $(sT+1)^{-1}$ and the phase detector (PD).

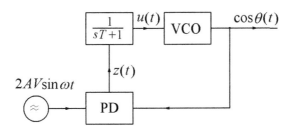

Fig. 3.1 The second order phase-locked loop.

The output signal $z(t)$ of PD depends on two 2π-periodic signals $U_{\text{inp}}(\omega t) = 2AV \sin \omega t$ and $U_{\text{out}}(\theta) = \cos \theta$. Usually one assumes

$$z(t) = 2AV \sin \omega t \, \cos \theta \equiv AV \big(-\sin \varphi + \sin(\varphi + 2\omega t) \big)$$

where $\varphi = \theta - \omega t$, but we suppose more general formula

$$z(t) = AV \big(-\sin \varphi + \varepsilon w(\varphi, \omega t) \big),$$

where ε is a positive parameter and $w(\varphi, \tau)$ is a 2π-periodic function of both arguments. This formula can be applied for some class of phase detectors which are not ideal multipliers.

We assume that $\sup\limits_{\varphi,\tau}|w(\varphi,\tau)| \leq 1$ and $\sup\limits_{\varphi,\tau}\left|\dfrac{\partial w(\varphi,\tau)}{\partial\varphi}\right| \leq 1$. Moreover, the mean value of the function $w(\varphi,\tau)$ with respect to τ is equal to zero for any fixed value of φ.

The phase-locked loop is described by the system of three equations

$$\frac{d\theta(t)}{dt} = \Omega\left(1 + \frac{u(t)}{V}\right),$$

$$T\frac{du(t)}{dt} + u(t) = z(t), \tag{3.1}$$

$$z(t) = AV\big(-\sin(\theta - \omega t) + \varepsilon w(\theta - \omega t, \omega t)\big),$$

for VCO, for the filter and for PD respectively.

In order to reduce the number of parameters we introduce new dimensionless variables

$$\tau = \omega t, \quad \varphi = \theta - \omega t, \quad x = \frac{u}{AV} \tag{3.2}$$

and parameters

$$\Delta = \frac{\Omega - \omega}{\Omega A}, \quad a = A\Omega T, \quad \nu = \omega T. \tag{3.3}$$

We get

$$\nu\frac{d\varphi}{d\tau} = a(x + \Delta),$$

$$\nu\frac{dx}{d\tau} = -(x + \sin\varphi) + \varepsilon w(\varphi,\tau). \tag{3.4}$$

In technical literature the time-varying term $w(\varphi,\omega t)$ is usually neglected and the autonomous system

$$\nu\frac{d\varphi}{d\tau} = a(x + \Delta),$$

$$\nu\frac{dx}{d\tau} = -(x + \sin\varphi) \tag{3.5}$$

is investigated. The system of equations (3.5) is called the *averaged system* with respect to (3.4). If the parameter ν takes a large value, then the solutions of the systems (3.4) and (3.5), starting from the same initial values, are close to each other in a large time interval. This is a consequence of general theorems on the *averaging method* (see Sec. 1.3) and this is intuitively obvious: an inert object which moves slowly is not sensitive to fast varying perturbations.

Let us notice an analogy between the PLL system described by equations (3.5) and a certain mechanical system. Eliminating the variable x from the equations (3.5) and changing the time scale $\tau_w = \frac{\sqrt{a}}{\nu}\tau = \sqrt{\frac{A\Omega}{T}}t$, we get

$$\frac{d^2\varphi}{d\tau_w^2} + \frac{1}{\sqrt{a}}\frac{d\varphi}{d\tau_w} + \sin\varphi = \Delta. \tag{3.6}$$

This is the equation of a pendulum with coefficient of viscous friction equal to $1/\sqrt{a}$ and forced by the constant torque Δ. The qualitative analysis of solutions of the equation (3.6) is apparently known since 1933 and it can be found in Ref. [2] for example. The equation (3.6) describes also the phenomena in Josephson's junction excited by d.c. source. A part of results presented in this chapter is taken from the paper [45].

3.2 Phase-plane portrait of the averaged system

The solutions of the equations (3.5) depend on two parameters a and Δ, where $a > 0$. The parameter ν changes the time scale only.

Replacement φ, x, Δ by $-\varphi$, $-x$, $-\Delta$ does not change the equation (3.5). If the functions $\varphi(\tau; \Delta_0)$, $x(\tau; \Delta_0)$ denote the solution of the equation (3.5) for a fixed positive value Δ_0 of the parameter Δ then the functions $-\varphi(\tau; -\Delta_0)$, $-x(\tau; -\Delta_0)$ fulfil the equations (3.5) for $\Delta = -\Delta_0 < 0$. So, it is sufficient to restrict the analysis to the case $\Delta \geq 0$.

From periodicity of the right hand side of the equations (3.5) with respect to φ we conclude that if the functions $\varphi(\tau)$, $x(\tau)$ fulfil these equations, then the functions $\varphi(\tau) + 2\pi$, $x(\tau)$ also fulfil them. It is convenient to assume, that $\varphi \bmod 2\pi$ is a cyclic variable, or to restrict the analysis to the range of $\varphi \in [-\pi, \pi)$.

3.2.1 The phase-plane trajectories

Let the pair of the functions $\varphi(\tau; \varphi_0, x_0)$, $x(\tau; \varphi_0, x_0)$ be such a solution of the equations (3.5) which satisfies the initial condition $\varphi(0; \varphi_0, x_0) = \varphi_0$, $x(0; \varphi_0, x_0) = x_0$. The graph described in the φx-plane by the parametric representation

$$\varphi = \varphi(\tau; \varphi_0, x_0), \quad x = x(\tau; \varphi_0, x_0), \quad \tau \in (-\infty, +\infty) \tag{3.7}$$

is called the *phase-plane trajectory* passing through the point (φ_0, x_0). Exactly one trajectory passes through each point (φ_0, x_0) of the plane and it is tangent to the vector

$$\big(a(x_0 + \Delta), \ -(x_0 + \sin\varphi_0) \big). \tag{3.8}$$

at this point. In an exceptional case, when the vector (3.8) is equal to zero, the phase-plane trajectory reduces to the point. It is called a *singular point*, and the solution with the initial value at this point is a constant function.

Phase-plane trajectories (excluding singular points) fulfil the differential equation

$$\frac{dx}{d\varphi} = -\frac{x + \sin\varphi}{a(x + \Delta)} \tag{3.9}$$

which arises from (3.5) after eliminating the time τ. If we know the trajectories we can conclude about some qualitative properties of solutions of the equations (3.5). Therefore, we begin by sketching this trajectories in the φx-plane or on the cylinder $S \times R$, where $\varphi \bmod 2\pi$ is a cyclic variable.

The locus of points where the slope of trajectories with respect to the φ-axis is equal to λ is called an *isocline* with the slope λ. Its equation is

$$(x + \sin\varphi) + \lambda a(x + \Delta) = 0. \tag{3.10}$$

In particular, the isoclines of the horizontal direction ($\lambda = 0$) and of the vertical direction ($\lambda = \infty$) take the forms

$$x = -\sin\varphi \quad \text{and} \quad x = -\Delta. \tag{3.11}$$

The isoclines (3.11) separate some domains in the φx-plane. In each of these domains the vectors tangent to the trajectories belong to one of the four quadrants of Cartesian's system of coordinates. These quadrants are determined only by the signs of the right-hand side of the equations (3.5). For example, if the right-hand side of the first equation (3.5) is positive and the right-hand side of the second equation (3.5) is negative at a point (φ, x) then $\varphi(\tau)$ increases and $x(\tau)$ decreases at this point. Consequently, the vector tangent to the trajectory belongs to the fourth quadrant. The isoclines (3.11) (dotted lines), the respective quadrants and trajectories (bold lines) are shown in Fig.3.3 and Fig.3.5. Arrows show the direction of movement of the point (φ, x) along trajectories, when time τ increases.

For any initial condition (φ_0, x_0) the phase-plane trajectory enters the domain $|x| \leq 1$ and stays there for ever. The further behavior of the trajectories depends essentially on whether $|\Delta| < 1$ or $|\Delta| > 1$. We shall consider both cases more accurately.

3.2.2 The case $|\Delta| > 1$. Phase-modulated output signals

Let $A = (-\frac{3}{2}\pi, -1)$ and by $B = (-\frac{1}{2}\pi, 1)$ denote two points in the φx-plane. Let us consider the domain (see Fig.3.2) bounded by two segments of trajectories A–$f(A)$, B–$f(B)$ and by two segments of the isocline $x = -\sin\varphi$ of horizontal directions A–B, $f(A)$–$f(B)$.

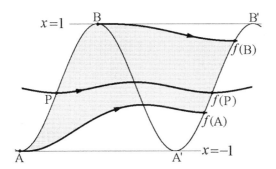

Fig. 3.2   Illustration to the proof on existence of a periodic trajectory $x = M(\varphi)$.

If $\varphi \bmod 2\pi$ is a cyclic variable, then the segment A'–B' is identical with A–B on the cylinder and, consequently, the segment $f(A)$–$f(B)$ is contained in A–B. Let $P = (\alpha, p)$ be an arbitrary point lying on the segment A–B, i.e. $\alpha = -\pi + \arcsin p$. The trajectory starting from the point P intersects the segment $f(A)$–$f(B)$ at the point $f(P) = (\alpha', p')$, where $\alpha' = \pi + \arcsin p'$. There exists on the segment A–B of the cylinder a fixed point $P^* = (\alpha^*, p^*)$ of the mapping f or, equivalently, such a point $P = (\alpha, p)$ that $p' = p$ and $\alpha' = 2\pi + \alpha$.

The trajectory passing through the point P^* is periodic. We denote it by

$$x = M(\varphi), \quad \text{where} \quad M(\varphi + 2\pi) = M(\varphi). \tag{3.12}$$

The function $x = M(\varphi)$ is a periodic solution of the equation (3.9) and therefore we have the identity

$$-a\Big(M(\varphi) + \Delta\Big)\frac{dM(\varphi)}{d\varphi} = M(\varphi) + \sin\varphi.$$

After integration this equality over the interval $[0, 2\pi]$ we get

$$-a\left(\frac{1}{2}M^2(2\pi) + \Delta M(2\pi) - \frac{1}{2}M^2(0) - \Delta M(0)\right) = \int_0^{2\pi}\big(M(\varphi) + \sin\varphi\big)d\varphi.$$

Since $M(2\pi) = M(0)$, we conclude that

$$\int_0^{2\pi} M(\varphi)d\varphi = 0. \tag{3.13}$$

The mean value of the function $M(\varphi)$ is equal to zero independently of the values of the parameters a and Δ. So, there exists exactly one periodic trajectory (in the opposite case, two periodic trajectories would have intersection points, which is impossible).

For $\Delta > 1$ the right-hand side of the equation (3.9) is a function of the class C^1 in the domain $x > -\Delta$ and

$$\frac{\partial}{\partial x}\left(-\frac{\sin\varphi + x}{a(x + \Delta)}\right) = -\frac{\Delta - \sin\varphi}{a(x + \Delta)^2} < 0. \tag{3.14}$$

By Theorem 2.3 the solution $x = M(\varphi)$ of the equation (3.9) is stable, i.e. the neighboring trajectories come close to the curve $x = M(\varphi)$ for $\varphi \to \infty$.

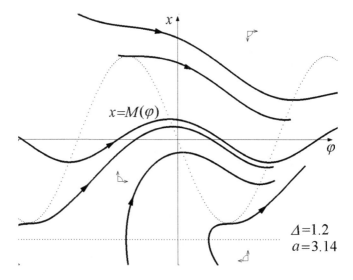

Fig. 3.3　Family of phase-plane trajectories (bold lines) and isoclines (dotted lines) of the system (3.5) for $\Delta > 1$.

The effective determination of the function $x = M(\varphi)$ demands numerical calculation. For large and small values of the parameter a we have asymptotic expressions

$$M(\varphi) = \frac{\cos\varphi}{a\Delta} + O\left(1/a^2\right) \tag{3.15}$$

and

$$M(\varphi) = -\sin\varphi + a(\Delta - \sin\varphi)\cos\varphi + O\left(a^2\right) \qquad (3.16)$$

which follow from Theorems 2.4 and 2.5 respectively. Let us note that the parameter a is inversely proportional to the cut-off frequency of the low-pass filter.

Finally, we conclude that for $\Delta > 1$ all trajectories tend to the periodic trajectory $x = M(\varphi)$ as $\varphi \to \infty$. An example of such a family of trajectories is shown in Fig.3.3. If $x > -\Delta$ then $\varphi(\tau)$ is an increasing function. So, if we neglect the transient state, then the phase-locked loop can be described by the equation

$$\frac{d\varphi}{dt} = A\Omega\left(\Delta + M(\varphi)\right). \qquad (3.17)$$

Basic properties of such equations are given in Section 2.2. The frequency of the output steady-state signal of PLL is periodically modulated and it has a mean value not equal to ω.

3.2.3   The case $|\Delta| < 1$. Hold-in region

Intersection points of isoclines are singular points. In the range $\varphi \in [0, 2\pi)$ the system of equations (3.5) has exactly two singular points

$$(\arcsin\Delta, -\Delta) \quad \text{and} \quad (\pi - \arcsin\Delta, -\Delta). \qquad (3.18)$$

To determine the character of these points we use the well known theorem (see Ref.[50], [25], [2]).

Theorem 3.1. *Let (u_0, v_0) be a singular point of the system of equations*

$$\frac{du}{d\tau} = P(u, v), \qquad \frac{dv}{d\tau} = Q(u, v), \qquad (3.19)$$

where the functions P, Q have continuous partial derivatives of the first order.

Let

$$J_0 = \begin{bmatrix} P'_u(u_0, v_0) & P'_v(u_0, v_0) \\ Q'_u(u_0, v_0) & Q'_v(u_0, v_0) \end{bmatrix} \qquad (3.20)$$

denote the Jacobian matrix at the singular point (u_0, v_0). Let $\det J_0$ and $\operatorname{Tr} J_0$ denote the determinant (Jacobian) and the trace of the matrix (3.20) respectively.

The character of a singular point depends on these parameters in the following way:

– *for* $\det J_0 < 0$ *the singular point is a saddle point,*
– *for* $0 < 4 \det J_0 < (\text{Tr } J_0)^2$ *the singular point is a node; stable if* $\text{Tr } J_0 < 0$ *and unstable if* $\text{Tr } J_0 > 0$,
– *for* $4 \det J_0 > (\text{Tr } J_0)^2$ *the singular point is a focus; stable if* $\text{Tr } J_0 < 0$ *and unstable if* $\text{Tr } J_0 > 0$,
 – *in other cases the character of a singular point depends on higher order derivatives of the functions P and Q at this point.*

A family of phase-plane trajectories (the phase-plane portrait) in a neighborhood of a singular point is shown in Fig.3.4 for a saddle point, node and focus respectively.

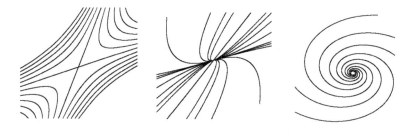

Fig. 3.4   Phase-plane portraits in a neighborhood of singular points: saddle point, node and focus.

For the system of equations (3.5) the matrix (3.20) takes the form

$$J_0(\varphi_0, x_0) = \frac{1}{\nu} \begin{bmatrix} 0 & a \\ -\cos \varphi_0 & -1 \end{bmatrix}. \tag{3.21}$$

Consequently, $\nu \text{Tr } J_0 = -1$ and $\nu^2 \det J_0 = a \cos \varphi_0 = \pm a \sqrt{1 - \Delta^2}$. The first singular point (3.18) is a stable node (for $4a\sqrt{1 - \Delta^2} < 1$) or a stable focus (for $4a\sqrt{1 - \Delta^2} > 1$). The second point (3.18) is a saddle point.

The domain of parameters a, Δ for which the system (3.5) has a stable fixed point is called the *hold-in region*.

Each saddle point has two pairs of *separatrices*, i.e. trajectories for which the saddle point is a limit point. We call the trajectories of one of these pairs *approaching* the saddle point (S_1 and S_3 in Fig.3.5), and the trajectories of the other pair *leaving* the saddle point (S_2 and S_4 in Fig.3.5). The two saddle points shown in Fig.3.5 become the same point on the cylinder where $\varphi \bmod 2\pi$ is a cyclic variable.

Let us now deal with the *attractive domain* of the stable singular point, i.e. the set of all initial points for which the solutions of the equations (3.5)

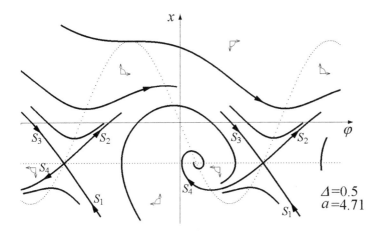

Fig. 3.5   Family of phase-plane trajectories (bold lines) and isoclines (dotted lines) of the system (3.5) for $0 < \Delta < 1$.

tend to the point $(\arcsin \Delta, -\Delta)$. The boundary of this domain consists of the separatrices S_1 and S_3 approaching the saddle point. The shape of the attractive domain (on the cylinder) depends on the mutual location of the separatrices S_2 and S_3. All three possibilities are shown in Fig.3.6.

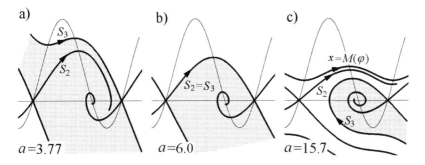

Fig. 3.6   Attractive domains of stable singular points for $\Delta = 0.5$ and for three values of the parameter a.

If the separatrix S_3 lies below S_2 (Fig.3.6c) then the attractive domain of the stable singular point is bounded from above. The trajectories which are not attracted to the stable point (and are not separatrices approaching the saddle point) tend to the periodic trajectory $x = M(\varphi)$, just like in the case $\Delta > 1$. The proof of the existence of the periodic trajectory in this case differs from the proof carried out for the case $\Delta > 1$ only in nonessential details.

If the separatrix S_3 lies above S_2 (Fig.3.6a) then the attractive domain of the stable singular point is not bounded. It includes all points of the cylinder with the exception of the separatrices S_1, S_3 approaching the saddle point. In this case the periodic trajectory does not exist. This happens in such a domain of parameters a, Δ which is called the *pull-in region*.

If the separatrices S_2 and S_3 overlap, the points lying below $S_2 \equiv S_3$ (with exception of points lying on S_1) belong to the attractive domain of the stable singular point. The points lying above the separatrix $S_2 \equiv S_3$ are attracted to this separatrix.

Fig.3.7 shows an $S \times R$ cylinder with graphs of the periodic trajectory and separatrices of the saddle point of the system (3.5) for selected values of parameters.

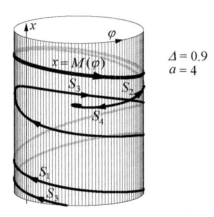

$\Delta = 0.9$
$a = 4$

Fig. 3.7   Cylinder with graphs of the periodic trajectory and separatrices of the saddle point, for $H(a) < \Delta < 1$.

With reference to the notion of attractive domain we have the following:

Comments. Suppose that for $t < 0$ the input signal is $2A_0 V \sin \omega_0 t$ and the system (3.5) is in the stable equilibrium point (φ_0, x_0), where $\varphi_0 = \arcsin \Delta_0$, $x_0 = -\Delta_0$ and $\Delta_0 = \frac{\Omega - \omega_0}{\Omega A_0}$. The output signal of PLL is $\cos(\omega_0 t + \varphi_0)$.

At the moment $t = 0$ the input signal changes and takes the value $2A_1 V \sin \omega_1 t$. For $t > 0$ dynamics is described by the equation (3.5) with new parameters $\nu_1 = \omega_1 T$, $\Delta_1 = \frac{\Omega - \omega_1}{\Omega A_1}$ and $a_1 = A_1 \Omega T$ which differs from $\nu_0 = \omega_0 T$, Δ_0 and $a_0 = A_0 \Omega T$. The initial value (φ_0, x_0) for the new equation is determined by the equilibrium point for the previous parameters.

For $t > 0$ trajectories of the system tend to new stable equilibrium point (φ_1, x_1), where $\varphi_1 = \arcsin \Delta_1$, $x_1 = -\Delta_1$ if and only if the initial point (φ_0, x_0) belongs to the attractive domain of the new system on the cylinder. It is possible that trajectories many times run round the cylinder before they come to a small neighborhood of the point (φ_1, x_1).

If (φ_0, x_0) is out of the attractive domain (of the system with parameters Δ_1, a_1) then synchronization disappears and the output signal of PLL does not tend to $\cos(\omega_1 t + \varphi_1)$ as $t \to \infty$.

3.2.4 *Boundary of pull-in region:* $S_2 \equiv S_3$

Let $x = M_s(\varphi)$ be the equation of the separatrix $S_2 \equiv S_3$ shown in Fig.3.6b. This separatrix satisfies the equation (3.9) and the boundary conditions

$$M_s(\varphi_0) = M_s(\varphi_0 + 2\pi) = -\Delta \quad \text{with} \quad \varphi_0 = -\pi - \arcsin \Delta, \qquad (3.22)$$

where $(\varphi_0, -\Delta)$ is the saddle point. Moreover, it is easy to prove that the equality

$$\int_{\varphi_0}^{\varphi_0 + 2\pi} M_s(\varphi) d\varphi = 0 \qquad (3.23)$$

holds (the proof is the same as in Section 3.2.2 for the periodic trajectory).

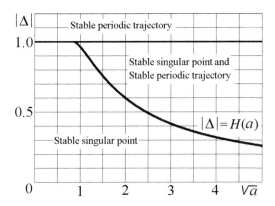

Fig. 3.8   Domains of the parameters $\sqrt{a}, \Delta$ for which the stable limit sets exist: stable periodic trajectory, stable singular point and both of these.

For a given value of a the separatrix $S_2 \equiv S_3$ exists only for $|\Delta| = H(a)$, where the graph of the function H have to be determined numerically. It is shown in Fig.3.8 in the plane of coordinates $\sqrt{a}, |\Delta|$.

For $a_0 = 0.70256...$ we have $H(a_0) = 1$ and for $a \in (a_0, \infty)$ the function H decreases from one to zero. Indeed, if the value of a increases then $|\frac{dx}{d\varphi}|$ decreases (the graph of the solution of the equation (3.9) becomes flat), and the conditions (3.22), (3.23) can be satisfied only when the value of $|\Delta|$ decreases. We will prove in Section 3.8 that for large values of the parameter a the asymptotic equality

$$H(a) \approx \frac{4}{\pi} \frac{1}{\sqrt{a}} \tag{3.24}$$

holds.

If $a > a_0$ and $H(a) < |\Delta| < 1$ then the system of the equations (3.5) has two stable limit sets: the stable periodic trajectory $x = M(\varphi)$ and the stable singular point $(\arcsin \Delta, -\Delta)$, which is either a focus or node. If $|\Delta| < H(a)$ then only the second stable limit set exists, i.e. the singular point.

3.2.5   The case $|\Delta| = 1$. Boundary of hold-in region

On the cylinder $S \times R$ (or in the plane in the range $\varphi \in [0, 2\pi]$) the system of the equations (3.5) has for $\Delta = 1$ exactly one singular point $(\frac{\pi}{2}, -1)$ of the saddle-node type. Let us move the origin of the coordinate system to the singular point, assuming $\varphi = \beta + \frac{\pi}{2}$, $x = \xi - 1$. For the new variables β, ξ the differential equation (3.9) of phase-plane trajectories is of the form

$$a\xi \frac{d\xi}{d\beta} = 2\sin^2 \frac{\beta}{2} - \xi. \tag{3.25}$$

The phase-plane trajectories (solutions of the equation (3.25)) in a neighborhood of the point $\beta = \xi = 0$ are shown in Fig.3.9. Among these trajectories it is necessary to distinguish two curves passing through the origin of the coordinate system. One of them, of the form

$$\xi = -\frac{1}{a}\beta - \frac{1}{4}\beta^2 + \frac{a}{24}\beta^3 + \frac{4 - 5a^2}{384}\beta^4 + \dots \tag{3.26}$$

determines the separatrices S_1 and S_3, and the second one, i.e.

$$\xi = 2\sin^2 \frac{\beta}{2} - \frac{a}{2}\beta^3 + \frac{5a^2}{4}\beta^4 + \dots \quad \text{where} \quad \beta > 0 \tag{3.27}$$

determines the separatrix S_2. It is easy to check that both functions fulfil the equation (3.25) with an error of the order $O(\beta^5)$. All trajectories

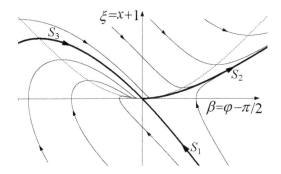

Fig. 3.9 Trajectories in a neighborhood of the singular point of the saddle-node type. Solutions of Eq. (3.25).

from left-hand side of the line (3.26) tend to the singular point and have a horizontal slope at this point.

If $a = a_0 \equiv 0.70256...$ then separatrix $S_2 \equiv S_3$ encircle the cylinder. If $a > a_0$ then separatrix S_2 tends to the periodic trajectory and if $a < a_0$ then it tends to the singular point. The global behavior of separatrices is similar to that shown in Fig.3.6. However, for $\Delta = 1$ two singular points (the saddle point and the node) join together to one point of the saddle-node type.

3.2.6 The filter with high cut-off frequency

If the low-pass filter has high cut-off frequency then T takes small value and a is small.

Let us change the time scale in the equations (3.5), assuming $\tau_s = \frac{a}{\nu}\tau = A\Omega t$. We get the system of equations

$$\frac{d\varphi}{d\tau_s} = x + \Delta, \qquad \frac{dx}{d\tau_s} = -\frac{x + \sin\varphi}{a} \qquad (3.28)$$

with a small parameter a. On the whole φx-plane, with exception of a small neighborhood of the line $x = -\sin\varphi$, the vector tangent to trajectory has the vertical component significantly bigger then the horizontal component. Therefore, the point (φ, x) moves quickly (with velocity $\left|\frac{dx}{d\tau_s}\right|$ of the order a^{-1}) to the line $x = -\sin\varphi$ along almost vertical trajectories and then it stays in a neighborhood of this line.

Two cases are possible.

$1°$ If $|\Delta| > 1$ then all trajectories approach the periodic trajectory

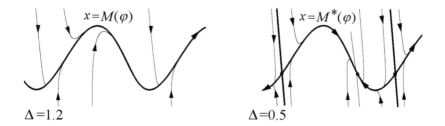

Fig. 3.10   Trajectories of the system (3.28) for small values of the parameter $a = 0.05$ and for two values of Δ. Bold lines denote the periodic trajectory or separatrices of the saddle point.

$x = M(\varphi)$ given by the formula (3.16). On this trajectory the horizontal component of velocity

$$\nu \frac{d\varphi}{d\tau} = a(\Delta - \sin\varphi) + a^2(\Delta - \sin\varphi)\cos\varphi + O(a^3) \qquad (3.29)$$

is comparable with that one where the filter is neglected (see Sec. 2.2.2).

2° If $|\Delta| < 1$ then there exist on the cylinder two singular points: saddle point and stable node. The line formed by separatrices leaving the saddle point and by singular points is a smooth closed curve $x = M^*(\varphi)$. All trajectories approach (quickly) this curve and tend (slowly) to the node.

Two pictures of phase-plane trajectories, each for a small value of the parameter a, are shown in Fig.3.10. The lines $x = M(\varphi)$ and $x = M^*(\varphi)$ take both an important role in the analysis of properties of the system (3.4) containing a time-varying component.

3.2.7   The filter with low cut-off frequency

If the low-pass filter has low cut-off frequency then T takes large value and a is large.

It is more convenient to use the system of equations

$$\frac{d\varphi}{d\tau_w} = v, \qquad \frac{dv}{d\tau_w} = \Delta - \sin\varphi - \frac{v}{\sqrt{a}} \qquad (3.30)$$

equivalent to (3.6), where $\tau_w = \sqrt{\frac{A\Omega}{T}}t$. The term $v/\sqrt{a}$ is treated as a small perturbation of the conservative system

$$\frac{d\varphi}{d\tau_w} = v, \qquad \frac{dv}{d\tau_w} = \Delta - \sin\varphi. \qquad (3.31)$$

The trajectories of the system (3.31) fulfil the differential equation $v\,dv =$
$= (\Delta - \sin\varphi)d\varphi$ which has the following one-parameter family of solutions:

$$\frac{1}{2}v^2 - \Delta\varphi - \cos\varphi = E, \tag{3.32}$$

where a constant value E is called the *energy* of the system. The family of
the lines (3.32) for several values of the parameter E is shown in Fig.3.11
for $\Delta = 0$, for $0 < \Delta < 1$ and for $\Delta > 1$.

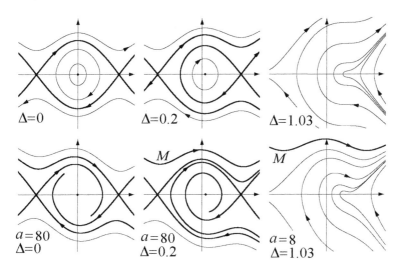

Fig. 3.11   Phase-plane trajectories of the conservative system (3.31) (upper roof) and
of the system (3.30) for large values of the parameter a (lower roof), both on the
φv-plane, where $v = \sqrt{a}(x + \Delta)$. The periodic trajectory is signed by M.

For a large value of the parameter a the trajectories of the system (3.30)
are close to the trajectories of (3.31) but only locally (i.e. in a short interval
of time), and are not close globally.

The differential equation of phase-plane trajectories of the system (3.30)
takes the form

$$v\,dv - (\Delta - \sin\varphi)d\varphi = -\frac{1}{\sqrt{a}}\,v\,d\varphi.$$

After integration we get

$$\frac{1}{2}v^2(\tau_w) - \Delta\varphi(\tau_w) - \cos\varphi(\tau_w) = E - E_d(\tau_w), \tag{3.33}$$

where

$$E_d(\tau_w) = \frac{1}{\sqrt{a}} \int_{\varphi(0)}^{\varphi(\tau_w)} v d\varphi = \frac{1}{\sqrt{a}} \int_0^{\tau_w} v^2(t) dt.$$

If time τ_w increases then the values of the function E_d slowly increase. This fact determines some qualitative properties of trajectories of the system (3.30). The trajectories are not the lines of constant energy, as in conservative system, but they cross these lines in direction of smaller values of energy. Some selected trajectories (especially the separatrices of the saddle point and periodic trajectories) of the system (3.30) are shown in Fig.3.11 for large values of the parameter a.

3.3   Perturbation of the phase difference $\varphi(\omega t)$

Now we investigate the effect of high-frequency component (which is neglected in averaged equations) on the phase difference $\varphi(\omega t)$ of the output signal

$$U_{\text{out}}\big(\theta(t)\big) = \cos\theta(t) = \cos\big(\omega t + \varphi(\omega t)\big).$$

If the averaged system (3.5) has an asymptotically stable equilibrium point (φ_0, x_0) then for all initial values from a neighborhood of this point the phase difference $\varphi(\omega t)$ tends to φ_0 as $t \to \infty$. It is not true for the system (3.4) with a high-frequency component $\varepsilon w(\varphi, \tau)$, particularly with $\varepsilon w(\varphi, \tau) = \sin(\varphi + 2\tau)$.

For various values of parameters the following cases are observed:

1) $\varphi(\tau)$ tends to a π-periodic or 2π-periodic function which takes values in a neighborhood of φ_0. The output signal of PLL has the same period as the input signal and its spectrum contains harmonic frequencies $k\omega$ for $k = 1, 2, 3, \dots$

2) $\varphi(\tau)$ tends to a $n\pi$-periodic function, where $n > 2$. In this case the output signal has also sub-harmonic frequencies $\frac{\omega}{2}$, $\frac{\omega}{4}$ and others.

3) $\varphi(\tau)$ is a bounded "chaotic" signal (where $\max_{\tau_1, \tau_2} |\varphi(\tau_1) - \varphi(\tau_2)| < 2\pi$) with a dense spectrum in the range of low frequencies. The output signal looks as a noised sinus-wave of the frequency ω.

4) no stable and bounded solution $\varphi(\tau)$ exists and no input-output synchronization takes place.

We shortly describe these cases with emphasis on the first case.

3.3.1   A basic theorem

We begin with a classical theorem about a small periodic perturbation of an equilibrium point. The formulation is restricted to the two-dimensional case, however the theorem is true (with minor changes) for the n-dimensional case (see Refs. [23], [14]).

Theorem 3.2. *Let us consider the differential equation*

$$\frac{dz}{dt} = F(z) + \varepsilon f(z,t), \qquad t \in (-\infty, +\infty), \tag{3.34}$$

where $z = \{u, v\}$ *and*

$$F(z) = \{F_1(u, v), F_2(u, v)\}, \qquad f(z,t) = \{f_1(u, v, t), f_2(u, v, t)\}.$$

Let z_0^* *denote an equilibrium point of the autonomous equation*

$$\frac{dz}{dt} = F(z), \tag{3.35}$$

i.e. $F(z_0^*) = 0$, *and let the real parts of the eigenvalues of the Jacobian matrix* $F'(z_0^*)$ *are different from zero.*

Let the functions F *and* f *be of the class* C^2 *with respect to* z *and let the function* f *be bounded for* z *belonging to a neighborhood of* z_0^* *and for* $t \in (-\infty, +\infty)$.

Under these assumptions there exists $\varepsilon_0 > 0$ *such that for all* $0 < \varepsilon < \varepsilon_0$ *the equation (3.34) has in a neighborhood of* z_0^* *exactly one solution* $z_\varepsilon^*(t)$ *bounded for* $t \in (-\infty, +\infty)$, *continuously dependent on* ε *and*

$$\sup_t |z_\varepsilon^*(t) - z_0^*| \to 0 \quad for \quad \varepsilon \to 0.$$

If the function f *is periodic with respect to* t, *then the solution* $z_\varepsilon^*(t)$ *is also periodic of the same period.*

If the singular point z_0^* *is a stable node or a stable focus then the solution* $z_\varepsilon^*(t)$ *attracts the neighboring solutions of the equation (3.34) for* $t \to +\infty$. *If* z_0^* *is an unstable node or an unstable focus then the solution* $z_\varepsilon^*(t)$ *repels the neighboring solutions (attracts for* $t \to -\infty$).

If z_0^* *is a saddle point then the solution* $z_\varepsilon^*(t)$ *is hyperbolic. It means that in a neighborhood of the line* $z_\varepsilon^*(t)$, $t \in (-\infty, +\infty)$ *there exist (in tuv-space) two surfaces* S_+ *and* S_- *which intersect along this line. The intersections of these surfaces with the plane* $t = const.$ *lie in uv-plane close to the separatrices of the saddle point* z_0^*. *If the initial condition* $z_\varepsilon(0)$ *lies on the surface* S_+ *(or on* S_-) *then the whole solution* $z_\varepsilon(t)$ *of the equation (3.34) remains on this surface and tends to* $z_\varepsilon^*(t)$ *for* $t \to +\infty$ *(for* $t \to -\infty$ *respectively).*

The proof is omitted.

It is not easy to obtain the periodic solution $z_\varepsilon^*(t)$ using standard numerical methods because we do not know the initial condition $z_\varepsilon^*(0)$ for it. If the periodic solution lies in a small neighborhood of a node or focus of the autonomous system (3.35), then we can obtain this solution as a steady--state by numerical integration of the differential equations (3.34) in a long interval of time with positive time direction, if the singular point is stable, or negative time direction, if it is unstable. If the periodic solution lies in a small neighborhood of a saddle point of the autonomous system then we are constrained to use more sophisticated methods.

3.3.2   *An approximate formula for periodic solutions*

Let us consider the system of equations

$$
\begin{aligned}
\nu \frac{d\varphi}{d\tau} &= a(x + \Delta), \\
\nu \frac{dx}{d\tau} &= -(x + \sin\varphi) + \sin(\varphi + 2\tau),
\end{aligned}
\tag{3.36}
$$

which describe the PLL with an ideal multiplier as a phase detector.

As a first approximation of a periodic solution of equations (3.36) we can take the solution of a special linear equation.

Let (φ_0, x_0) denote a singular point of the averaged system, i.e. $\sin\varphi_0 = -x_0 = \Delta$. The singular point is a saddle point if $\cos\varphi_0 < 0$ and a stable node or stable focus if $\cos\varphi_0 > 0$. The averaged system linearized in a neighborhood of its singular point and excited by the high-frequency term $\sin(\varphi_0 + 2\tau)$ takes the form

$$
\begin{aligned}
\nu \frac{d\beta}{d\tau} &= a\xi, \\
\nu \frac{d\xi}{d\tau} &= -\beta\cos\varphi_0 - \xi + \sin(\varphi_0 + 2\tau),
\end{aligned}
\tag{3.37}
$$

where $\beta = \varphi - \varphi_0$, $\xi = x - x_0$.

The periodic solution of the equations (3.37), expressed by functions with natural parameters A, ω, Ω (instead of $a = A\Omega T$ and $\nu = \omega T$), is given by

$$
\begin{aligned}
\beta(\tau) &= -\tfrac{1}{2}A\tfrac{\Omega}{\omega}|E|\cos(2\tau + \varphi_0 + \arg E), \\
\xi(\tau) &= |E|\sin(2\tau + \varphi_0 + \arg E),
\end{aligned}
\tag{3.38}
$$

where

$$|E| = \frac{1}{\sqrt{1 + (\frac{A}{2}\frac{\Omega}{\omega}\cos\varphi_0 - 2\omega T)^2}},$$ (3.39)

$$\arg E = \arctan(\frac{A}{2}\frac{\Omega}{\omega}\cos\varphi_0 - 2\omega T).$$

The "amplitude" $|E|$ depends on cut-off frequency of the filter. Evidently $|E| \leq 1$ and $|E|$ tends to zero as $\nu = \omega T$ tends to infinity. Moreover, for unstable periodic solution in a neighborhood of saddle point, the value of $|E|$ is smaller then for stable solution in a neighborhood of node or focus (which follows from the sign of $\cos\varphi_0$).

3.3.3 Numerical experiments

If ν is sufficiently large then in a neighborhood of the point (φ_0, x_0) the equation (3.36) has exactly one periodic solution and its period is equal to π. However, if ν is not sufficiently large then the equation (3.36) has usually several solutions of long periods or an infinite number of bounded "chaotic" solutions. It is illustrated by numerical experiments.

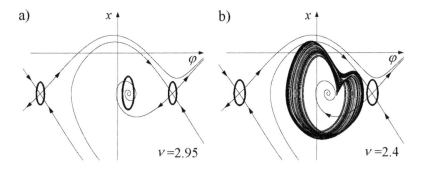

Fig. 3.12   Separatrices of saddle points of the autonomous system (3.5) and the projection (a) of periodic and (b) of "chaotic" solutions of Eq. (3.36) (bold lines) on the φx-plane.

For fixed parameters $a = 8$ and $\Delta = 0.5$ the separatrices of the saddle point of averaged system are shown in Fig.3.12. For $\nu > 2.945$ the system (3.36) has π-periodic solutions in a neighborhood of singular points. The projections of these solutions are shown in Fig.3.12a for $\nu = 2.95$. If $2.55 < \nu < 2.94$ then the system (3.36) has 2π-periodic stable solution in a neighborhood of the focus. If ν is a bit smaller then in the place of a stable periodic solution we have a stable "chaotic" solution (bounded, non-periodic). The projection of such a solution is shown in Fig.3.12b for $\nu = 2.4$.

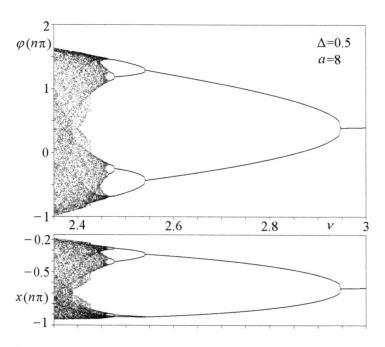

Fig. 3.13   The set of values $\{\varphi(n\pi), x(n\pi)\}$ of the steady-state solutions of Eq. (3.36) at the points $n\pi$ for $n = 70, \ldots, 120$ versus the parameter ν.

The following numerical experiment illustrates the influence of parameter ν on the stable periodic solution of the equation (3.36) in a neighborhood of the focus of the averaged equation. For 1000 values of ν from the interval (2.35, 3.00) we calculate the sequence $\{\varphi(n\pi), x(n\pi)\}$ of values of a solution of the equation (3.36) at the points $\tau = n\pi$ for $n = 0, \ldots, 120$. The initial values are given at the focus $\varphi(0) = \arcsin \Delta$, $x(0) = -\Delta$. The graph of sequences $\{\varphi(n\pi), x(n\pi)\}$ for $n = 70, \ldots, 120$ versus ν is shown in Fig.3.13. If for a fixed value of ν we observe N values of φ (and N values of x) then the steady-state periodic solution is of period $N\pi$. The stable solutions of periods $\pi, 2\pi, 4\pi, \ldots$ and non-periodic solutions (or periodic of very long periods) are observed while ν decreases from 3 to 2.35. The theory of this phenomena will be given in Section 4.7.

If $\nu < 2.34$ then no stable bounded solution exists in a neighborhood of the focus of averaged equation.

3.4 Stable integral manifold

Usually, dynamics of a second order PLL system is more complicated then of the first order. However, if the second order system has a globally stable integral manifold then it behaves as the first order system. In this section we introduce the notions and basic theorems on integral manifolds. This theory will be applied to the equations of PLL systems in the next section.

Let us consider a motion described by the differential equation

$$\frac{dz}{dt} = F(z,t), \quad t \in (-\infty, \infty), \tag{3.40}$$

where $z = \{z_1, \ldots, z_n\}$ and $F(z,t) = \{F_1(z,t), \ldots, F_n(z,t)\}$.

If $z(t)$ is a solution of the equation (3.40) then the set $\{(z(t),t) : t \in R\}$ is called the graph of this solution in $R^n \times R$ space.

3.4.1 *The basic notions and motivations*

In some cases if $t \to \infty$ then the graphs of solutions become close to some smooth manifold $\mathcal{M}_k$ of dimension $k + 1$, where $1 \leq k < n$. Moreover, if a point $(z(t_0), t_0)$ belongs to $\mathcal{M}_k$ then the whole graph $\{(z(t),t) : t \in (-\infty, +\infty)\}$ also belongs to $\mathcal{M}_k$. The manifold is made up of whole graphs of solutions. If a graph does not belong to $\mathcal{M}_k$ then it has no common point with $\mathcal{M}_k$. Such a manifold is called a k-dimensional stable integral manifold of the equation (3.40).

Not every differential equation has a stable integral manifold, but if such a manifold exists then it is very useful for the study of the motion. Since graphs of solutions tend to $\mathcal{M}_k$, one may accept that after a long time the motion will take place on the manifold $\mathcal{M}_k$. It will then be sufficient to use an equation of the motion in k-dimensional space. The existence of the stable integral manifold enables us to reduce the order of the differential equation which describes the motion. Moreover, the integral manifold is usually not sensitive to small perturbations of the right-hand side of the equations even though these perturbations significantly change the graphs of particular solutions.

We shall present a simple example of the integral manifold. Let us imagine the motion of a ball in a neighborhood of the wavy lake surface. No matter whether the ball is initially above or under the water, it quickly falls or emerges from the water and remains on the surface. The free motion of the ball is described by the sixth order equation (3 space coordinates and 3 velocities), but on the water surface the fourth order equation (2 space coordinates and 2 velocities tangent to the surface) is sufficient. So, the surface of water determines the 4-dimensional stable integral manifold in a 6-dimensional space of the free motion.

Now the formal definition of an integral manifold of the equation (3.40) will be given (see Ref. [24]).

Let $\mathcal{M}_k$ be such a set of points (z, t) of the space $R^n \times R$ which has the parametric representation

$$\mathcal{M}_k = \big\{ (z, t) : z = h(u, t), \ u \in R^k, \ t \in R \big\}, \quad 1 \leq k \leq n - 1, \quad (3.41)$$

where the bounded function h is a homeomorphism of $R^k \times R$ onto $\mathcal{M}_k$. The function h has continuous and bounded partial derivatives with respect to $u = \{u_1, ..., u_k\}$, and the matrix $\dfrac{\partial(h_1, h_2, \ldots, h_n)}{\partial(u_1, u_2, \ldots, u_k)}$ is of the rank k.

Definition 3.1. The set (3.41) is called the integral manifold of the equation (3.40) if from $(z_0, t_0) \in \mathcal{M}_k$ it follows that the graph of the solution $z(t; z_0, t_0)$ of the equation (3.40) with the boundary condition $z(t_0; z_0, t_0) = z_0$ lies entirely on $\mathcal{M}_k$, i.e.

$$\big(z(t; z_0, t_0), t \big) \in \mathcal{M}_k \quad \text{for } t \in (-\infty, +\infty). \quad (3.42)$$

The manifold $\mathcal{M}_k$ is called periodic (almost periodic) if the function $h(u, t)$ is periodic (almost periodic) with respect to t. The manifold $\mathcal{M}_k$ is called cylindrical if $h(u, t) = h_c(u)$ is independent of t.

Let dist $[p, \mathcal{M}_k]$ denote the distance of a point p from the manifold $\mathcal{M}_k$ in the space $R^n \times R$.

Definition 3.2. The manifold $\mathcal{M}_k$ is called stable if for any $\varepsilon > 0$ there exists $\eta > 0$ such that from dist $[(z_0, t_0), \mathcal{M}_k] < \eta$ it follows that dist $\big[\big(z(t; z_0, t_0), t \big), \mathcal{M}_k \big] < \varepsilon$ for $t > t_0$ and, moreover, if

$$\text{dist} \big[\big(z(t; z_0, t_0), t \big), \mathcal{M}_k \big] \to 0 \quad \text{for } t \to +\infty. \quad (3.43)$$

The stable manifold $\mathcal{M}_k$ is called globally stable, if the condition (3.43) is satisfied for any initial point $(z_0, t_0) \in R^n \times R$.

3.4.2   *An equation of the second order*

Let us consider the system of differential equations

$$\frac{dx}{dt} = -Cx + P(x, \varphi, t)$$
$$\frac{d\varphi}{dt} = f(x, \varphi, t)$$

(3.44)

in the set

$$D = \{(x, \varphi, t) : |x| \leq \delta, \ \varphi \in R, \ t \in R\}, \tag{3.45}$$

where C and δ are fixed positive numbers. The function P is bounded $\sup_D |P(x, \varphi, t)| \leq m$. Moreover, both functions P and f satisfy in D the following Lipschitz conditions:

$$|P(x_1, \varphi_1, t) - P(x_2, \varphi_2, t)| \leq \lambda_a |x_1 - x_2| + \lambda_b |\varphi_1 - \varphi_2|, \tag{3.46}$$

$$|f(x_1, \varphi_1, t) - f(x_2, \varphi_2, t)| \leq \lambda_c |x_1 - x_2| + \lambda_d |\varphi_1 - \varphi_2|, \tag{3.47}$$

and they are continuous with respect to t.

Let $B_{\delta\mu}$ denote the set of continuous functions $h(\varphi, t)$ which satisfy the conditions

$$\sup_{\varphi, t} |h(\varphi, t)| \leq \delta, \qquad \sup_{\varphi_1, \varphi_2, t} \left| \frac{h(\varphi_1, t) - h(\varphi_2, t)}{\varphi_1 - \varphi_2} \right| \leq \mu. \tag{3.48}$$

Theorem 3.3. *If*

$$m \leq \delta\, C \quad \text{and} \quad \mu\lambda_a + \lambda_b + \mu^2\lambda_c + \mu\lambda_d \leq \mu\, C, \tag{3.49}$$

then, in the set $B_{\delta\mu}$, there exists exactly one function $h^*(\varphi, t)$ for which

$$\{(x, \varphi, t) : \ x = h^*(\varphi, t), \ \varphi \in R, \ t \in R\} \tag{3.50}$$

is the integral manifold of the system (3.44). This manifold is stable. If the right-hand side of the equations (3.44) is periodic with respect to t, or φ, or both variables, then the function $h^*(\varphi, t)$ has the same property.

If the assumptions of Theorem 3.3 are satisfied, then solutions $x(t), \varphi(t)$ of the equations (3.44) tend to the surface $x = h^*(\varphi, t)$ for $t \to \infty$ provided that the initial points are in the set D. After a long time we can admit that $x(t) \approx h^*(\varphi(t), t)$, where the function $\varphi(t)$ satisfies the differential equation of the first order

$$\frac{d\varphi}{dt} = f\big(h^*(\varphi, t), \varphi, t\big). \tag{3.51}$$

Even if we do not know the equation of the manifold, but we know that it exists and it is stable, then we can get some qualitative properties of steady-state solutions of the system (3.44).

The function $h^*(\varphi, t)$ which describes the integral manifold is a solution of the partial differential equation

$$Ch^* - P(h^*, \varphi, t) + \frac{\partial h^*}{\partial \varphi} f(h^*, \varphi, t) + \frac{\partial h^*}{\partial t} = 0. \qquad (3.52)$$

Indeed, if the solution $x(t), \varphi(t)$ of the equations (3.44) lies on the manifold then $x(t) - h^*\big(\varphi(t), t\big) \equiv 0$ for every t. By differentiation of this expression and the use of the equations (3.44) we get (3.52).

3.4.3   *Proof of Theorem 3.3*

The proof will be divided into 6 steps.

$1°$ estimation of solutions of the second equation (3.44) for an arbitrary fixed function $x = h(\varphi, t)$,

$2°$ construction of functions $h_n(\varphi, t)$ which approximate an integral manifold,

$3°$ periodicity of the functions $h_n(\varphi, t)$,

$4°$ convergence of the sequence $\{h_n(\varphi, t)\}$,

$5°$ proof that the limit function determines the integral manifold,

$6°$ stability of the manifold.

Initial points of differential equations are marked by the upper index zero.

$1°$ Let the functions $h_1(\varphi, t)$ and $h_2(\varphi, t)$ belong to the set $B_{\delta\mu}$. Let $\varphi_1(t)$ and $\varphi_2(t)$ be the solutions of the Cauchy problems

$$\frac{d\varphi_1}{dt} = f\big(h_1(\varphi_1, t), \varphi_1, t\big), \quad \varphi_1(t^0) = \alpha_1, \qquad (3.53)$$

$$\frac{d\varphi_2}{dt} = f\big(h_2(\varphi_2, t), \varphi_2, t\big), \quad \varphi_2(t^0) = \alpha_2, \qquad (3.54)$$

respectively.

Lemma 3.1. *The following estimation:*

$$|\varphi_1(t) - \varphi_2(t)| \leq |\alpha_1 - \alpha_2| \, e^{s|t-t^0|} + \lambda_c \frac{e^{s|t-t^0|} - 1}{s} \, ||h_1 - h_2||, \qquad (3.55)$$

holds, where

$$s = \lambda_d + \mu\lambda_c, \quad ||h|| = \sup_{\varphi, t} |h(\varphi, t)|$$

Proof. From the inequality

$$|h_1(\varphi_1, t) - h_2(\varphi_2, t)| \le |h_1(\varphi_1, t) - h_1(\varphi_2, t)| + \tag{3.56}$$
$$+ |h_1(\varphi_2, t) - h_2(\varphi_2, t)| \le \mu|\varphi_1 - \varphi_2| + ||h_1 - h_2||$$

it follows that the absolute value of the difference of right-hand sides of the equations (3.53) and (3.54) is bounded from above by

$$(\lambda_d + \mu\lambda_c)|\varphi_1 - \varphi_2| + \lambda_c ||h_1 - h_2||.$$

The function $z(t) = |\varphi_1(t) - \varphi_2(t)|$ fulfils the differential inequalities

$$-sz(t) - \lambda_c||h_1 - h_2|| \le \frac{dz(t)}{dt} \le sz(t) + \lambda_c||h_1 - h_2||, \quad z(t^0) = |\alpha_1 - \alpha_2|.$$

Solving these inequalities separately for $t > t^0$ and $t < t^0$ we get (3.55). $\square$

2° We will form some sequence $\{h_n(\varphi, t)\}$, $n = 1, 2, ...$ of functions which approximate the integral manifold. The sequence will be defined recurrently. Let us admit that we already know $h_n(\varphi, t)$. We form the next term $h_{n+1}(\varphi, t)$ of the sequence in the following way.

First, we determine the solution $\varphi = \varphi_{hn}(t; \varphi^0, t^0)$ of the Cauchy problem

$$\frac{d\varphi}{dt} = f\big(h_n(\varphi, t), \varphi, t\big), \quad \varphi(t^0; \varphi^0, t^0) = \varphi^0. \tag{3.57}$$

Next, we introduce the notation

$$p_n(t; \varphi^0, t^0) = P\Big(h_n\big(\varphi_{hn}(t; \varphi^0, t^0), t\big), \varphi_{hn}(t; \varphi^0, t^0), t\Big) \tag{3.58}$$

and instead of the first equation (3.44) we write the linear differential equation

$$\frac{dx}{dt} = -Cx + p_n(t; \varphi^0, t^0). \tag{3.59}$$

In the set of functions bounded on a whole time axis the equation (3.59) has exactly one solution

$$x_n(t; \varphi^0, t^0) = \int_0^\infty e^{-C\tau} p_n(t - \tau; \varphi^0, t^0)d\tau \equiv \int_{-\infty}^t e^{-C(t-\tau)} p_n(\tau; \varphi^0, t^0)d\tau.$$

This solution is used to construct the operation $\mathcal{F}$ which maps $h_n(\varphi, t)$ into $h_{n+1}(\varphi, t)$. We admit that for every initial point (φ^0, t^0), there is $h_{n+1}(\varphi^0, t^0) = x_n(t^0; \varphi^0, t^0)$ or, in a more explicit form,

$$h_{n+1}(\varphi, t) = \int_0^\infty e^{-C\tau} p_n(t - \tau; \varphi, t) d\tau \equiv \int_{-\infty}^t e^{-C(t-\tau)} p_n(\tau; \varphi, t) d\tau, \quad (3.60)$$

where the function p_n depends on h_n by (3.58) and by the solution of the equation (3.57).

3° The periodicity of the functions h_n can be proved by induction.

Let P and f be 2π-periodic functions of φ, and $h_n(\varphi, t) = h_n(\varphi + 2\pi, t)$. Hence, the solution of the equation (3.57) satisfies the equality $\varphi_{hn}(t; \varphi^0 + 2\pi, t^0) = \varphi_{hn}(t; \varphi^0, t^0) + 2\pi$, the function (3.58) is periodic $p_n(t; \varphi^0 + 2\pi, t^0) = p_n(t; \varphi^0, t^0)$ and the function (3.60) is also periodic $h_{n+1}(\varphi, t) = h_{n+1}(\varphi + 2\pi, t)$. Evidently, if $h_0(\varphi, t)$ is a 2π-periodic function of φ, then all terms of sequence $\{h_n(\varphi, t)\}$ and its limit function (if it exists) are also 2π-periodic.

Now, let P and f be T-periodic functions of t, and $h_n(\varphi, t) = h_n(\varphi, t + T)$. Hence, the solution of the equation (3.57) and the function (3.58) satisfy equalities $\varphi_{hn}(t; \varphi^0, t^0) = \varphi_{hn}(t + T; \varphi^0, t^0 + T)$ and, consequently, $p_n(t; \varphi^0, t^0) = p_n(t + T; \varphi^0, t^0 + T)$. Now it is easy to check that the function (3.60) is periodic $h_{n+1}(\varphi, t) = h_{n+1}(\varphi, t + T)$. Finally, if $h_0(\varphi, t)$ is a T-periodic function of t, then all terms of sequence $\{h_n(\varphi, t)\}$ and its limit function (if it exists) are also T-periodic functions of t.

4° In the abbreviated notation we write $h_{n+1} = \mathcal{F}(h_n)$ in place of (3.60). We will prove that for arbitrary $h_0 \in B_{\delta\mu}$ the sequence $\{h_n\}$ tends to the fixed point $h^* \in B_{\delta\mu}$ of the operation $\mathcal{F}$.

Lemma 3.2. *If the inequalities* (3.49) *are satisfied then the operation* $\mathcal{F}$ *maps the set* $B_{\delta\mu}$ *into itself.*

Proof.   Let $\mathcal{F}(h) = g$, where $h = h(\varphi, t)$ and $g = g(\varphi, t)$. If $h \in B_{\delta\mu}$, then

$$|P(h, \varphi, t)| \le m \quad \text{and} \quad |g(\varphi, t)| \le m \int_0^\infty e^{-C\tau} d\tau = \frac{m}{C} \le \delta.$$

According to (3.46) we have the estimation

$$|g(\alpha_1, t) - g(\alpha_2, t)| \le \int_0^\infty e^{-C\tau} (\lambda_b + \mu\lambda_a) |\varphi_h(t - \tau; \alpha_1, t) - \varphi_h(t - \tau; \alpha_2, t)| d\tau,$$

where $\varphi = \varphi_h(t; \alpha, t^0)$ is the solution of the Cauchy problem

$$\frac{d\varphi}{dt} = f\big(h(\varphi, t), \varphi, t\big), \quad \varphi(t^0; \alpha, t^0) = \alpha,$$

and it satisfies the assumptions of Lemma 3.1. Thus, we have the estimate

$$|\varphi_h(t - \tau; \alpha_1, t) - \varphi_h(t - \tau; \alpha_2, t)| \le |\alpha_1 - \alpha_2| e^{s|\tau|},$$

where $s = \lambda_d + \mu \lambda_c$. It follows that

$$|g(\alpha_1, t) - g(\alpha_2, t)| \le (\lambda_b + \mu \lambda_a) \int_0^\infty e^{-(C-s)\tau} d\tau \, |\alpha_1 - \alpha_2| =$$

$$= \frac{\lambda_b + \mu \lambda_a}{C - s} |\alpha_1 - \alpha_2| \le \mu |\alpha_1 - \alpha_2|.$$

The last inequality is a consequence of (3.49). So, the function $g(\alpha, t)$ belongs to the set $B_{\delta\mu}$. □

Lemma 3.3. *The operation $\mathcal{F}$ defined in the set $B_{\delta\mu}$ satisfies the Lipschitz condition*

$$||\mathcal{F}(h_1) - \mathcal{F}(h_2)|| \le k ||h_1 - h_2|| \tag{3.61}$$

with the constant

$$k = \frac{1}{C}\left(\lambda_a + \lambda_c \frac{\lambda_b + \mu \lambda_a}{C - (\lambda_d + \mu \lambda_c)}\right)$$

which is less than one.

Proof.   From (3.47) and (3.56) it follows that

$$\left|P\big(h_1(\varphi_1, t), \varphi_1, t\big) - P\big(h_2(\varphi_2, t), \varphi_2, t\big)\right| \le (\lambda_b + \mu\lambda_a)|\varphi_1 - \varphi_2| + \lambda_a ||h_1 - h_2||.$$

Hence,

$$||\mathcal{F}h_1 - \mathcal{F}h_2|| \le$$

$$\le \sup_{\varphi, t} \int_0^\infty e^{-C\tau}\Big((\lambda_b + \mu\lambda_a)|\varphi_{h1}(t - \tau; \varphi, t) - \varphi_{h2}(t - \tau; \varphi, t)| + \lambda_a ||h_1 - h_2||\Big) d\tau.$$

The difference $|\varphi_{h1} - \varphi_{h2}|$ has been estimated in Lemma 3.1. Using (3.55) for $\alpha_1 = \alpha_2$, we get

$$||\mathcal{F}h_1 - \mathcal{F}h_2|| \le ||h_1 - h_2|| \int_0^\infty e^{-C\tau}\left((\lambda_b + \mu\lambda_a)\lambda_c \frac{e^{s\tau} - 1}{s} + \lambda_a\right) d\tau = k||h_1 - h_2||.$$

It is easy to check that from (3.49) it follows that $k < 1$. □

The set $B_{\delta\mu}$ with the metric defined by

$$\rho(h_1, h_2) = ||h_1 - h_2|| = \sup_{\varphi,t} |h_1(\varphi, t) - h_2(\varphi, t)| \qquad (3.62)$$

is a complete (and compact) set. By Lemmas 3.2 and 3.3 $\mathcal{F}$ is a contracting operation in this space and the equation $h = \mathcal{F}(h)$ has exactly one solution h^* in the set $B_{\delta\mu}$. Moreover, for arbitrary $h_0 \in B_{\delta\mu}$ the sequence $\{h_n\}$, where $h_{n+1} = \mathcal{F}(h_n)$, is convergent to h^*.

5° The limit function $h^*(\varphi, t)$ of the sequence $\{h_n(\varphi, t)\}$ has the following property. For an arbitrary initial point (φ^0, t^0) the solution $\varphi(t)$ of the Cauchy problem

$$\frac{d\varphi}{dt} = f\big(h^*(\varphi, t), \varphi, t\big), \qquad \varphi(t^0; \varphi^0, t^0) = \varphi^0$$

satisfies the integral equation of the manifold

$$h^*\big(\varphi(t), t\big) = \int_{-\infty}^{t} e^{-C(t-\tau)} P\Big(h^*\big(\varphi(\tau), \tau\big), \varphi(\tau), \tau\Big) d\tau.$$

Consequently, the function $x(t) = h^*\big(\varphi(t), t\big)$ is bounded for $t \in (-\infty, +\infty)$ and it satisfies the differential equation

$$\frac{dx}{dt} = -Cx + P\big(x, \varphi(t), t\big).$$

It means that the pair of functions $\varphi(t)$ and $x(t) = h^*\big(\varphi(t), t\big)$ is such a solution of equations (3.44) that its graph lies on the surface described by equation $x = h^*(\varphi, t)$. Thus, the set (3.50) is an integral manifold.

6° We will prove that the integral manifold is stable. Let $x(t)$, $\varphi(t)$ be an arbitrary solution of equations (3.44) lying in the domain D for $t \geq 0$ and not lying on the integral manifold. Let us consider the function

$$z(t) = x(t) - h^*\big(\varphi(t), t\big), \qquad (3.63)$$

which represents the deviation of solution $x(t)$, $\varphi(t)$ from the surface $x = h^*(\varphi, t)$ measured at the moment t along the line parallel to the x-axis. The function $z(t)$ preserves the sign. Assume that $z(t) > 0$ (if $z(t) < 0$ then the proof is similar). The derivative of the function $z(t)$ is

$$\frac{dz(t)}{dt} = -Cx(t) + P\big(x(t), \varphi(t), t\big) - \frac{\partial h^*}{\partial \varphi} f\big(x(t), \varphi(t), t\big) - \frac{\partial h^*}{\partial t}. \qquad (3.64)$$

The function $h^*(\varphi, t)$ satisfies the partial differential equation (3.52) at each point (φ, t). Let us admit $\varphi = \varphi(t)$. Combining (3.64) with (3.52) we get

$$\frac{dz(t)}{dt} = -Cz(t) + P\big(x(t), \varphi(t), t\big) - P\big(h^*(\varphi(t), t), \varphi(t), t\big) +$$
$$-\frac{\partial h^*}{\partial \varphi}\Big(f\big(x(t), \varphi(t), t\big) - f\big(h^*(\varphi(t), t), \varphi(t), t\big)\Big).$$

Using the Lipschitz conditions (3.46), (3.47) and the fact that $z(t) > 0$ we get

$$\frac{dz(t)}{dt} \leq -Cz(t) + (\lambda_a + \mu\lambda_c)z(t).$$

From (3.49) it follows that $\gamma = C - \lambda_a - \mu\lambda_c$ is a positive number. Hence

$$z(t) \leq z(0)e^{-\gamma t} \quad \text{for } t > 0,$$

and the integral manifold is stable and exponentially attracting.
It finishes the proof of Theorem 3.3.

3.4.4   *Uniqueness of the manifold*

Let the functions $P(x, \varphi, t)$, $f(x, \varphi, t)$ be periodic with respect to φ and t with periods 2π and T respectively.

Theorem 3.4. *If the expression*

$$W(x, \varphi, t) = -C + \frac{\partial P(x, \varphi, t)}{\partial x} + \frac{\partial f(x, \varphi, t)}{\partial \varphi} \qquad (3.65)$$

does not change its sign in the domain $D = \{(x, \varphi, t) : a \leq x \leq b, \ \varphi \in R, \ t \in R\}$ *then at most one integral manifold* $x = h^*(\varphi, t)$ *exists in* D.

Proof.   Suppose that two manifolds

$$x = h_1^*(\varphi, t), \quad x = h_2^*(\varphi, t), \qquad (3.66)$$

periodic with respect to both variables, exist in D. They cannot intersect. Let us admit that $a \leq h_1^*(\varphi, t) < h_2^*(\varphi, t) \leq b$. Let us consider the solid (Fig.3.14)

$$D^* = \{(x, \varphi, t) : h_1^*(\varphi, t) \leq x \leq h_2^*(\varphi, t), \ 0 \leq \varphi \leq 2\pi, \ 0 \leq t \leq T\} \quad (3.67)$$

and the vector field

$$\vec{F}(x, \varphi, t) = \{-Cx + P(x, \varphi, t), \ f(x, \varphi, t), \ 1\} \qquad (3.68)$$

which is periodic with respect to φ and t.

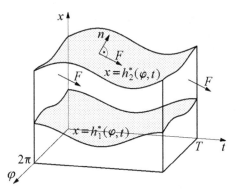

Fig. 3.14   Illustration to the proof of theorem on uniqueness of integral manifold.

According to the divergence theorem we have

$$\iiint_{D^*} \operatorname{div}\vec{F}\,dv = \iint_{\partial D^*} \vec{F}\cdot\vec{n}\,ds, \tag{3.69}$$

where ∂D^* is boundary of D^*, and $\vec{n}$ is the unit vector normal to ∂D^* and pointing out of D^*.

The solid D^* has three pairs of faces. The surface integral over the union of two faces $t = 0$ and $t = T$ is equal to zero because $\vec{F}(x, \varphi, 0) = = \vec{F}(x, \varphi, T)$ and exterior normal vectors have opposite directions on these faces. Similarly, the surface integral over the union of faces $\varphi = 0$ and $\varphi = 2\pi$ is also equal to zero. In every point of the faces described by equations

$$-x + h_i^*(\varphi, t) = 0, \quad i = 1, 2, \quad \varphi \in [0, 2\pi], \quad t \in [0, T]$$

(i.e. on the manifolds) the vector $\vec{n} = \left\{-1, \dfrac{\partial h^*}{\partial \varphi}, \dfrac{\partial h^*}{\partial t}\right\}$, which is normal to the manifold, is perpendicular to the vector (3.68) according to (3.52). So, the scalar product $\vec{F} \cdot \vec{n}$ is equal to zero in every point of manifolds (3.66). Obviously, the right-hand side of (3.69) is equal to zero, but the left-hand side of (3.69) is not equal to zero because the function $\operatorname{div}\vec{F}(x, \varphi, t) \equiv \equiv W(x, \varphi, t)$ preserves its sign. We conclude that the supposition that there exist two periodic manifolds (3.66) leads us to contradiction.   □

3.5   The PLL system reducible to the first order one

In this section sufficient conditions will be given, for which the PLL system described by the equation (3.4) can be reduced to a first order system. The first order equation designates dynamics on the stable periodic integral manifold

$$\mathcal{M} = \{(x, \varphi, \tau) : x = h^*(\varphi, \tau), \ \varphi \in R, \ \tau \in R\}. \tag{3.70}$$

Such manifold will be called the *integral surface* $x = h^*(\varphi, \tau)$, in short.

The autonomous system (3.5) has the integral surface $x = M(\varphi)$ (independent of τ) for $|\Delta| > 1$ and also for $H(a) < |\Delta| < 1$, where $a > a_0 = 0.70256...$ The properties of this surface were given in Sections 3.2.2 and 3.2.4. Moreover, for small values of parameter a and for $|\Delta| < 1$ there exists the integral surface $x = M^*(\varphi)$ formed by the graphs of such solutions that their projections on φx-plane are separatrices leaving the saddle point (see Sec. 3.2.6). We will prove that in neighborhoods of these surfaces the system (3.4) with $\varepsilon \neq 0$ has also the integral surfaces (3.70).

Proposition 3.1. *The system* (3.4) *has at most one integral surface.*

Proof.   The divergence of the vector field defined by the right-hand sides of equations (3.4) is $W(x, \varphi, \tau) = -1$ and, by Theorem 3.4, we have at most one integral surface (3.70). □

3.5.1   *Small values of parameter* $a = A\Omega T$

It will be proved that for small values of a there exists an integral surface which reduces to $x = M(\varphi)$ or to $x = M^*(\varphi)$ as $\varepsilon \to 0$.

Theorem 3.5. *If*

$$4a(1 + \varepsilon) \leq 1 \tag{3.71}$$

then the system (3.4) *has a stable integral surface* $x = h^*(\varphi, \tau)$ *which is* 2π*-periodic with respect to both variables. Moreover*

$$|h^*(\varphi, \tau)| \leq 1 + \varepsilon \quad and \quad |h^*(\varphi_1, \tau) - h^*(\varphi_2, \tau)| \leq 2(1 + \varepsilon)|\varphi_1 - \varphi_2|.$$

Proof.   Comparing the systems of equations (3.4) and (3.44) one can easily calculate that the parameters defined by (3.46)–(3.48) and multiplied by ν take the following values:

$$\nu C = 1, \quad \nu m = 1 + \varepsilon, \quad \nu \lambda_a = 0, \quad \nu \lambda_b = 1 + \varepsilon, \quad \nu \lambda_c = a, \quad \nu \lambda_d = 0$$

If the inequalities (3.49) hold i.e. if the conditions

$$1 + \varepsilon \le \delta \quad \text{and} \quad (1 + \varepsilon) + \mu^2 a \le \mu \tag{3.72}$$

are satisfied, then, in the set $B_{\delta\mu}$, there exists exactly one integral surface $x = h^*(\varphi, \tau)$.

It is easy to check that the inequalities (3.72) have positive solutions with respect to δ and μ if and only if $4a(1 + \varepsilon) \le 1$. One of these solutions is: $\delta = 1 + \varepsilon$, $\mu = 2(1 + \varepsilon)$. It finishes the proof. $\qquad\square$

Example 3.1. The autonomous system $\frac{dx}{d\tau} = -x - \sin\varphi$, $\frac{d\varphi}{d\tau} = ax$ has a stable integral surface for $a \le 0.25$, which follows from Theorem 3.5. This surface $x = M^*(\varphi)$ is formed by separatrices leaving the saddle points and by singular points, i.e. nodes $(0, 2\pi n)$ and saddle points $(0, (2n+1)\pi)$. For $a > 0.25$ the surface does not exist because nodes are replaced by focuses.

If the output signal $z(t) = 2AV \sin\omega t \cos\theta$ of phase detector PD is the product of its two input signals, then $\varepsilon w(\varphi, \tau) = \sin(\varphi + 2\tau)$. In this case the PLL system (3.36) has an integral surface $x = h^*(\varphi, \tau)$ for $a \le 0.125$. This surface is a 2π-periodic (with respect to φ and τ) solution of the partial differential equation

$$\nu \frac{\partial h^*}{\partial \tau} + a(h^* + \Delta) \frac{\partial h^*}{\partial \varphi} + h^* + \sin\varphi - \sin(\varphi + 2\tau) = 0 \tag{3.73}$$

If values of the parameter a are sufficiently small, then the solution can be obtained as a formal power series

$$h^*(\varphi, \tau) = h_0(\varphi, \tau) + a h_1(\varphi, \tau) + a^2 h_2(\varphi, \tau) + \dots \tag{3.74}$$

Substituting h^* by this formula in (3.73) and comparing terms with the same power of a, we get the system of linear equations

$$\nu \frac{\partial h_0}{\partial \tau} + h_0 = -\sin\varphi + \sin(\varphi + 2\tau),$$

$$\nu \frac{\partial h_1}{\partial \tau} + h_1 = -(h_0 + \Delta) \frac{\partial h_0}{\partial \varphi},$$

$$\cdots \cdots \cdots \cdots \cdots \cdots$$

the periodic solutions of which can be successively found. Hence we get the solution

$$h^*(\varphi, \tau) = -\sin\varphi + \frac{\sin(\varphi + 2\tau - \alpha)}{\sqrt{1 + 4\nu^2}} + a\big((\Delta - \sin\varphi)\cos\varphi + O(\nu^{-2})\big) + O(a^2).$$

where $\alpha = \arctan 2\nu$ and where $O(\nu^{-2})$ replaces several terms which are as small as ν^{-2}.

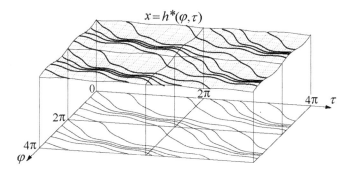

Fig. 3.15   Graphs of solutions of Eq. (3.36) on the integral surface (3.70) and their projections on $\varphi\tau$-plane.

If an initial value (x^0, φ^0, τ^0) for the system (3.36) lies on the surface $x = h^*(\varphi, \tau)$, then $\varphi(\tau)$ satisfies the first order differential equation

$$\nu\frac{d\varphi}{d\tau} = a\big(h^*(\varphi, \tau) + \Delta\big), \qquad \varphi(\tau^0) = \varphi^0. \tag{3.75}$$

Graphs of a family of solutions of equations (3.36) lying on the surface $x = h^*(\varphi, \tau)$ and their projections on $\varphi\tau$-plane i.e. the graphs of solutions of equation (3.75) are shown in Fig.3.15.

3.5.2   A neighborhood of the trajectory $x = M(\varphi)$

Let us consider the equations (3.4) in such a region of parameters a, Δ, where the autonomous system (with $\varepsilon = 0$) has a stable cylindrical integral manifold

$$\{(x, \varphi, \tau) : x = M(\varphi),\ \varphi \in R,\ \tau \in R\}, \quad \text{where } M(\varphi+2\pi) = M(\varphi), \tag{3.76}$$

which was investigated in Sections 3.2.2 and 3.2.4.

Theorem 3.6. *Let the autonomous system (3.5) have the periodic trajectory $x = M(\varphi)$. For $\varepsilon > 0$ sufficiently small, there exists a stable integral surface $x = h^*(\varphi, \tau)$ of the system (3.4). It is 2π-periodic with respect to φ and τ, it depends on ε and*

$$\sup_{\varphi,\tau} |h^*(\varphi, \tau) - M(\varphi)| \to 0 \quad \text{for } \varepsilon \to 0.$$

Proof.   The periodic trajectory satisfies the inequality $M(\varphi) + \Delta \neq 0$. Let us change variables in equations (3.4) assuming

$$\beta = \frac{1}{a} \int_0^\varphi \frac{d\alpha}{M(\alpha) + \Delta}, \quad z = \big(M(\varphi) + \Delta\big)\big(x - M(\varphi)\big). \tag{3.77}$$

The mapping $(\varphi, x) \rightarrow \big(\beta(\varphi), z(\varphi, x)\big)$ is a diffeomorphism, and there exists a mapping $(\beta, z) \rightarrow \big(\varphi(\beta), x(\beta, z)\big)$ inverse to (3.77). For the new system of coordinates βz, the equations (3.4) take the form

$$\nu \frac{dz}{d\tau} = -z - Q(\beta)z^2 + \varepsilon R(\beta, \tau),$$

$$\nu \frac{d\beta}{d\tau} = 1 + F(\beta)z, \tag{3.78}$$

where

$$Q(\beta) = \frac{M(\varphi) + \sin\varphi}{(M(\varphi) + \Delta)^3}, \quad F(\beta) = \frac{1}{(M(\varphi) + \Delta)^2},$$

$$R(\beta, \tau) = \big(M(\varphi) + \Delta\big)w(\varphi, \tau), \quad \varphi = \varphi(\beta).$$

The functions Q, F, R are T_0-periodic with respect to β, where $T_0 = \beta(2\pi)$ (and 2π-periodic with respect to φ). Moreover, they have a bounded and continuous derivative with respect to β because

$$\frac{dM(\varphi)}{d\varphi} = -\frac{1}{a} \frac{M(\varphi) + \sin\varphi}{M(\varphi) + \Delta} \quad \text{and} \quad \frac{d\varphi}{d\beta} = a\big(M(\varphi) + \Delta\big)$$

are continuous and bounded functions of β.

Theorem 3.3 will be used to prove the existence of the integral manifold

$$\{(z, \beta, \tau) : z = h_*(\beta, \tau), \ \beta \in R, \ \tau \in R\} \tag{3.79}$$

of equations (3.78) in the set $B_{\delta\mu}$ of functions $h(\beta, \tau)$ which are bounded by δ and which satisfy the Lipschitz conditions with respect to β with the constant μ.

Comparing the systems of equations (3.78) and (3.44) one can easily calculate that the parameters occurring in the assumptions of Theorem 3.3 depend on ε and δ in the following way:

$$\nu C = 1, \quad \nu m = A_1\delta^2 + A_2\varepsilon,$$

$$\nu\lambda_a = A_3\delta, \quad \nu\lambda_b = A_4\delta^2 + A_5\varepsilon, \quad \nu\lambda_c = A_6, \quad \nu\lambda_d = A_7\delta,$$

where $A_i, i = 1, 2, ...7$ are some positive numbers. If the inequalities (3.49) hold i.e. if the conditions

$$A_1\delta^2 + A_2\varepsilon \leq \delta,$$

$$A_3\mu\delta + A_4\delta^2 + A_5\varepsilon + A_6\mu^2 + A_7\mu\delta \leq \mu \tag{3.80}$$

are satisfied, then, in the set $B_{\delta\mu}$, there exists exactly one integral surface (3.79). It is easy to check that if ε is sufficiently small then the inequalities (3.80) have positive solutions with respect to δ and μ, and some of these solutions tend to zero for ε tending to zero. Since (3.77) is a diffeomorphism, then the equations (3.4) also have the integral surface (3.70), where

$$h^*(\varphi,\tau) = M(\varphi) + \frac{h_*\big(\beta(\varphi),\tau\big)}{M(\varphi) + \Delta}. \tag{3.81}$$

It finishes the proof.   □

3.6 Homoclinic structures

In averaged system (3.5) the boundary of attractive domain of a stable singular point consist of two separatrices approaching the saddle point (see Section 3.2.3). In the system (3.4) with a periodic high-frequency term the separatrices are replaced by selected invariant lines of the Poincaré mapping which can look very complicated. Consequently, we observe strange boundaries of attractive domains.

In this section we give two-dimensional version of the Poincaré mapping, we introduce the notion of homoclinic and heteroclinic trajectories and prove the Mielnikov theorem on the existence of such trajectories. In the next section we indicate how these techniques may be used to the equations of phase-locked loops.

3.6.1 *The Poincaré mapping*

Let us consider the system of equations

$$\frac{dx}{d\tau} = F_1(x,\varphi,\tau), \quad \frac{d\varphi}{d\tau} = F_2(x,\varphi,\tau), \tag{3.82}$$

where F_1 and F_2 are 2π-periodic functions with respect to both φ and τ. Let

$$x = x(\tau; x_0,\varphi_0), \quad \varphi = \varphi(\tau; x_0,\varphi_0)$$

denote the solution of the system (3.82) with the initial condition (x_0,φ_0), i.e. $x(0; x_0,\varphi_0) = x_0$, $\varphi(0; x_0,\varphi_0) = \varphi_0$.

Definition 3.3. A function P which maps an initial value (x_0,φ_0) onto the value (x_1,φ_1) of the solution at the moment $\tau = 2\pi$, i.e.

$$x_1 = x(2\pi; x_0,\varphi_0), \quad \varphi_1 = \varphi(2\pi; x_0,\varphi_0),$$

is called the Poincaré mapping.

We will use the notation $(x_1, \varphi_1) = P(x_0, \varphi_0)$. If the functions F_1 and F_2 have bounded and continuous partial derivatives with respect to x and φ, then P is a diffeomorphism of the $x\varphi$-plane onto itself.

Let P^m denote the m-th iterate of the Poincaré mapping. In particular, P^{-1} is the mapping inverse to P, and P^0 is the identity. It is easy to check that the equality

$$P^m(x_0, \varphi_0) = \big(x(2\pi m; x_0, \varphi_0), \; \varphi(2\pi m; x_0, \varphi_0)\big) \qquad (3.83)$$

holds for each integer m.

The sequence $\{P^m(x_0, \varphi_0)\}$, for all integers m, is called the *trajectory* of the point (x_0, φ_0). It is called the *positive* or *negative semi-trajectory* if the integers m takes positive or negative values respectively.

Periodicity of the functions F_1 and F_2 with respect to φ leads to

$$x(\tau; x_0, \varphi_0 + 2\pi) = x(\tau; x_0, \varphi_0), \quad \varphi(\tau; x_0, \varphi_0 + 2\pi) = \varphi(\tau; x_0, \varphi_0) + 2\pi,$$

and consequently

$$P(x, \varphi + 2\pi) = P(x, \varphi) + (0, 2\pi). \qquad (3.84)$$

It is convenient to treat P as a mapping of the cylinder $R \times S$ onto itself, where $\varphi \bmod 2\pi$ is a cyclic variable.

Definition 3.4. The point (x^*, φ^*) is called a periodic point of the type n/m of the mapping P, if there exist integers n and m such that

$$P^m(x^*, \varphi^*) = (x^*, \varphi^* + 2\pi n) \qquad (3.85)$$

and $P^k(x^*, \varphi^*) \neq (x^*, \varphi^* + 2\pi l)$ for $k = 1, 2, \ldots, m-1$ and an integer l.

If (x^*, φ^*) is a periodic point of the type n/m of the mapping P, then the solution $x(\tau) = x(\tau; x^*, \varphi^*)$, $\varphi(\tau) = \varphi(\tau; x^*, \varphi^*)$ of equations (3.82) satisfy the equalities

$$x(\tau + 2\pi m) = x(\tau), \quad \varphi(\tau + 2\pi m) = \varphi(\tau) + 2\pi n$$

for every τ. The trajectory $\{P^k(x^*, \varphi^*)\}$ (for all integers k) is a periodic sequence of period m on the cylinder.

Definition 3.5. A subset I of the cylinder $R \times S$ is called an invariant set of the Poincaré mapping if $P(I) = I$.

An example of invariant set I_m is a periodic trajectory of the type n/m. The set I_m has exactly m points. Another example of invariant set is the line

$$L_h = \{(x, \varphi) : x = h^*(\varphi, 0), \; \varphi \in S\}, \qquad (3.86)$$

where $x = h^*(x, \tau)$ is the integral surface discussed in Section 3.5.

If the right-hand sides of equations (3.82) do not depend on τ, then each whole trajectory of the autonomous system is an invariant line of the Poincaré mapping.

3.6.2 Invariant lines of hyperbolic fixed points

Let us describe the equations (3.82) in vectorial notation

$$\frac{dz}{d\tau} = F(z, \tau). \tag{3.87}$$

Periodicity of the vector-function $F = (F_1, F_2)$ with respect to the second variable of the vector $z = (x, \varphi)$ is not essential in this section.

Let the equation (3.87) have 2π-periodic solution $z_p(\tau)$. The point $z^* = z_p(0)$ is a fixed point of the Poincaré mapping in the $x\varphi$-plane. In a neighborhood of z^* the Poincaré mapping can be approximated by its linear part

$$P(z) = z^* + P'(z^*)(z - z^*) + O(|z - z^*|^2), \tag{3.88}$$

where the matrix $P'(z^*)$ (derivative of the mapping P at the point z^*) is equal to the value $W(2\pi)$ of the solution $W(\tau)$ of the linear differential equation

$$\frac{dW(\tau)}{d\tau} = F_z'(z_p(\tau), \tau) W(\tau), \quad W(0) = I. \tag{3.89}$$

Indeed, if $z(\tau; z_0)$ satisfies the equation (3.87) and the initial condition $z_0 = z(0; z_0)$, then we have the identity

$$\frac{\partial z(\tau, z_0)}{\partial \tau} \equiv F(z(\tau, z_0), \tau)$$

for all τ and z_0. Hence, the derivative $W(\tau, z_0) = \dfrac{\partial z(\tau, z_0)}{\partial z_0}$ satisfies the differential equation

$$\frac{\partial W(\tau, z_0)}{\partial \tau} = F_z'(z(\tau, z_0), \tau) W(\tau, z_0)$$

and the initial condition $W(0, z_0) = I$, where I is a unit-matrix. From $P(z_0) = z(2\pi, z_0)$ it follows that $P'(z_0) = W(2\pi, z_0)$. For the fixed value $z_0 = z^*$ the function $W(\tau) = W(\tau, z^*)$ satisfies the equation (3.89), and $P'(z^*) = W(2\pi)$.

Let the eigenvalues λ_+, λ_- of the matrix $P'(z^*)$ be real, positive and $\lambda_+ < 1 < \lambda_-$. Then in a neighborhood of the point z^* there exist two invariant lines L_+ and L_- which intersect at z^* and are tangent at this point to respective eigenvectors of the matrix $P'(z^*)$. If $z_0 \in L_+$ then $P^m(z_0) \in L_+$ and $P^m(z_0) \to z^*$ for $m \to +\infty$, with the rate of geometric progression with the ratio λ_+. Similarly, if $z_0 \in L_-$ then $P^m(z_0) \in L_-$ and $P^m(z_0) \to z^*$ for $m \to -\infty$, with the rate of geometric progression with the ratio $1/\lambda_-$. Such a fixed point z^* of the Poincaré mapping is called the *hyperbolic fixed point*. The lines L_+, L_- are called the *attracted* and the *repelled* invariant line of the point z^* respectively.

If we know a segment S_L of the invariant line L_+ (or L_-) with the end-points z_0 and $P(z_0)$, then we can extend it to the whole line by the formula $L_+ = \bigcup_{m=-\infty}^{+\infty} P^m(S_L)$.

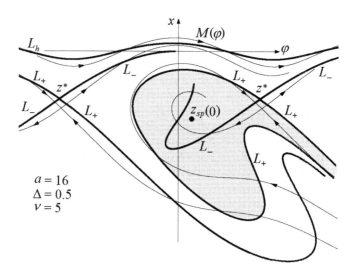

$a = 16$
$\Delta = 0.5$
$v = 5$

Fig. 3.16 Invariant lines L_+, L_- and L_h of the system (3.36) (bold lines). The periodic trajectory $x = M(\varphi)$ and separatrices of the saddle points of the averaged system (thin lines).

If the system of equations is of the form (3.34) instead of (3.87), then for small values of the parameter ε, the point z^* lies near the saddle point of the autonomous equation (3.35), and the invariant lines L_+, L_- are identical with the intersections of the surfaces S_+, S_- with the plane $\tau = 0$ (see Theorem 3.2).

An example of invariant lines L_+, L_- and L_h of the system (3.36) is shown in Fig.3.16. In the same figure the separatrices of the saddle point and the periodic trajectory $x = M(\varphi)$ of the autonomous system are marked by thin lines. In this case the line L_+ is the boundary between the set of initial values (grey domain in the figure) which are attracted to the stable periodic solution $z_{sp}(\tau)$ and these which are attracted to the stable integral surface $x = h^*(\varphi, \tau)$.

3.6.3 Heteroclinic and homoclinic trajectories

Invariant lines of the Poincaré mapping of a periodically time-varying system play a similar role to that of phase-plane trajectories of an autonomous system. However, they do not preserve all properties of these trajectories. Two different invariant lines can cross each other.

Let z^* and z^{**} be two hyperbolic fixed points of the Poincaré mapping of the system (3.87).

Definition 3.6. The sequence $\{P^m(z)\}$ is called a heteroclinic trajectory of the points z^*, z^{**} if $P^m(z) \to z^*$ as $m \to +\infty$ and $P^m(z) \to z^{**}$ as $m \to -\infty$.

The sequence $\{P^m(z)\}$, where $z \neq z^*$, is called a homoclinic trajectory of the hyperbolic invariant point z^* if $P^m(z) \to z^*$ as $m \to \pm\infty$.

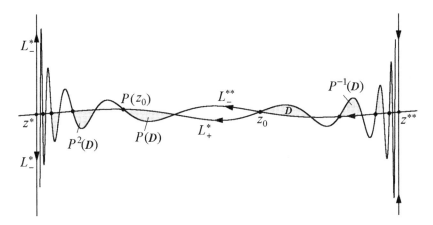

Fig. 3.17   Scheme of invariant lines L_+^* and L_-^{**}, and two heteroclinic trajectories.

The heteroclinic trajectory of the points z^*, z^{**} lies on the attracted invariant line L_+^* of the point z^* as well as on the repelled invariant line L_-^{**} of the point z^{**}. The homoclinic trajectory of the point z^* lies on the attracted invariant line L_+^* as well as on the repelled invariant line L_-^* of the point z^*.

An example of the invariant lines L_+^* and L_-^{**} of two hyperbolic fixed points z^* and z^{**} is shown in Fig.3.17. The points of intersection of these lines form two homoclinic trajectories (one of them is marked by black dots). The grey domain D, which is bounded by the segments of two lines L_+^* and L_-^{**}, is mapped successively by the Poincaré mapping onto the domains $P^m(D)$, which are marked in grey, for $m = -3, -2, -1, 0, 1, 2, 3, 4$.

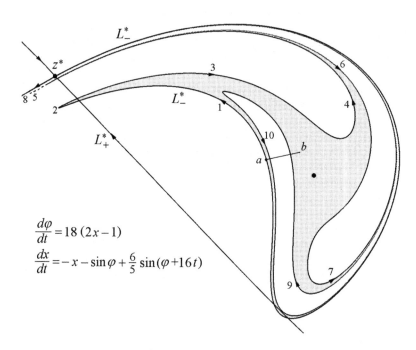

$$\frac{d\varphi}{dt} = 18\,(2x-1)$$

$$\frac{dx}{dt} = -x - \sin\varphi + \frac{6}{5}\,\sin(\varphi + 16\,t)$$

Fig. 3.18   Segment of the invariant line L_-^* (marked by successive numbers $1, 2, ..., 10$) in the case where there exists a homoclinic trajectory. The domain which lies on the right of the line L_-^* is marked in grey.

In a neighborhood of the point z^* the invariant line L_-^{**} containing a heteroclinic trajectory has a characteristic shape which is similar to the graph of the function

$$y(x) = \frac{1}{x^p}\Big(\delta + \cos(2\pi \ln x / \ln \lambda_+)\Big), \quad \text{where} \quad p = -\frac{\ln \lambda_-}{\ln \lambda_+} > 0, \quad (3.90)$$

with a parameter $\delta \in (-1, +1)$. The system of co-ordinates has been chosen in such a way that the axes of variables x, y have directions of eigenvectors of the matrix $P'(z^*)$. There exist two sequences of points $(x_k, y_k) = (C_1(\lambda_+)^k, 0)$ and $(x_k, y_k) = (C_2(\lambda_+)^k, 0)$ which converge to zero as $k \to \infty$. Between the successive values x_k there occur alternately the values of local maxima and minima of the function (3.90). The sequence of extremal values $y_{k\,\max}$ (or $y_{k\,\min}$) tends to $+\infty$ (or to $-\infty$) as $k \to \infty$, with the rate of geometric progression with the ratio λ_-. In every neighborhood of the point z^* the invariant line L_-^* is the limiting line of an infinite number of segments of the line L_-^{**}.

If there exists a homoclinic trajectory, i.e. if $z^* = z^{**}$ then the line L_-^* "winds onto itself in an oscillatory way" as is shown in Fig.3.18. Let $a\text{–}b$ be a segment transversal to the line L_-^* at an arbitrary point $a \in L_-^*$ and directed to the right of L_-^*. This segment has an infinite number of intersection points with the line L_-^* (in both directions). Particularly, in every neighborhood of the point z^* the line L_+^* has an infinite number of intersection points with the line L_-^*.

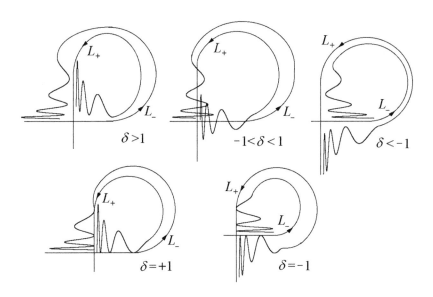

Fig. 3.19   Scheme of bifurcation of a homoclinic trajectory.

Usually the Poincaré mapping depends on a parameter. Changing the value of this parameter we can control the rise and disappearance of homoclinic or heteroclinic trajectories. The homoclinic trajectories arise in the way shown in Fig.3.19. The lines L_- and L_+ come close to each other and, for a bifurcation value of the parameter, become tangent at an infinite number of points forming the homoclinic trajectory. After crossing over the bifurcation value of the parameter, the lines L_- and L_+ intersect transversally along two homoclinic trajectories. The successive sketches in Fig.3.19 correspond to five values of parameter δ in formula (3.90). The bifurcation values for which a homoclinic trajectory appears are $\delta = 1$ and $\delta = -1$.

3.6.4   Melnikov's theorem

Let us consider the differential equation

$$\frac{dz}{dt} = F(z) + \varepsilon f(z,t), \quad t \in (-\infty, +\infty) \tag{3.91}$$

with a small positive parameter ε, where

$$z = \{u,v\}, \quad F(z) = \{F_1(u,v), F_2(u,v)\}, \quad f(z,t) = \{f_1(u,v,t), f_2(u,v,t)\}.$$

Similarly to Section 3.3.1, we assume that the functions F and f are of the class C^2 with respect to z, and f is a 2π-periodic function of t bounded for all z and t.

Let the autonomous equation (3.91) for $\varepsilon = 0$ has two saddle points z^*, z^{**} and a separatrix from one to the other. The solutions of the autonomous system lying on this separatrix form a one-parameter family of functions denoted by $z_0(t-t_0)$. For a fixed value of the parameter t_0 we have $z_0(t - t_0) \to z^*$ as $t \to -\infty$ and $z_0(t - t_0) \to z^{**}$ as $t \to +\infty$.

From Theorem 3.2 it follows that for a sufficiently small ε, in neighborhoods of the points z^*, z^{**}, the system (3.91) has 2π-periodic isolated hyperbolic solutions $z_\varepsilon^*(t)$, $z_\varepsilon^{**}(t)$. There exists a one-parameter family of solutions $z_\varepsilon^-(t; t_0)$ lying on the surface S_-^*, i.e. such solutions that $|z_\varepsilon^-(t; t_0) - z_\varepsilon^*(t)| \to 0$ as $t \to -\infty$. The intersection of the surface S_-^* with the plane $t = t_0$ forms the invariant line L_-^* repelled from the point $z_\varepsilon^*(t_0)$. The invariant line L_-^* has a parametric representation $z = z_\varepsilon^-(t_0, t_0)$.

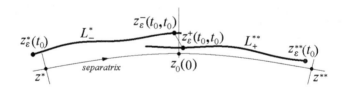

Fig. 3.20   Illustration to the notion of Melnikov's function.

Similarly, there exists a one-parameter family of solutions $z_\varepsilon^+(t; t_0)$ lying on the surface S_+^{**}, i.e. such solutions that $|z_\varepsilon^+(t; t_0) - z_\varepsilon^{**}(t)| \to 0$ as $t \to +\infty$. The intersection of the surface S_+^{**} with the plane $t = t_0$ forms the invariant line L_+^{**} attracted to the point $z_\varepsilon^{**}(t_0)$. The invariant line L_+^{**} has a parametric representation $z = z_\varepsilon^+(t_0, t_0)$.

The existence of a heteroclinic trajectory of points $z_\varepsilon^*(t_0)$, $z_\varepsilon^{**}(t_0)$ is equivalent to the existence of intersection points of the lines L_-^* and L_+^{**}.

Let us introduce the following notations:

$$F \wedge f = \det \begin{bmatrix} F_1 & F_2 \\ f_1 & f_2 \end{bmatrix} \equiv |F| \cdot |f| \sin(\angle F, f), \qquad (3.92)$$

$$\operatorname{div} F(z) = \frac{\partial F_1(u, v)}{\partial u} + \frac{\partial F_2(u, v)}{\partial v}, \qquad (3.93)$$

$$G(t) = \exp\left(-\int_0^t \operatorname{div} F(z_0(\tau)) d\tau \right), \qquad (3.94)$$

where $z_0(t)$ is the solution of the autonomous equation lying on the separatrix.

The periodic function

$$M(t_0) = \int_{-\infty}^{+\infty} F(z_0(t)) \wedge f(z_0(t), t + t_0) G(t) dt \qquad (3.95)$$

is called Melnikov's function (see Ref. [20]). It will be proved that $M(t_0)$ is a certain measure of the relative distance between the points $z_\varepsilon^-(t_0, t_0)$ and $z_\varepsilon^+(t_0, t_0)$ lying on the lines L_-^* and L_+^{**} respectively.

Theorem 3.7. *If Melnikov's function $M(t_0)$ changes its sign then for $\varepsilon > 0$ sufficiently small the lines L_-^* and L_+^{**} intersect transversally, and there exists a heteroclinic trajectory. If the function $M(t_0)$ takes positive values or negative values only then the lines L_-^* and L_+^{**} do not have common points for small values of ε.*

Proof. The proof will be divided into three parts:
1° construction of a geometric form of the Melnikov function $M(t_0)$,
2° extraction of a differential equation for two components of $M(t_0)$,
3° derivation of an analytic form of $M(t_0)$ by solving the above mentioned equations.

1° Let t_0 be a fixed parameter. The solutions of the system (3.91) lying on the surfaces S_-^* and S_+^{**} are represented by

$$\begin{aligned}
z_\varepsilon^-(t, t_0) &= z_0(t - t_0) + \varepsilon z_1^-(t, t_0) + O(\varepsilon^2) \quad \text{for } t \in (-\infty, t_0], \\
z_\varepsilon^+(t, t_0) &= z_0(t - t_0) + \varepsilon z_1^+(t, t_0) + O(\varepsilon^2) \quad \text{for } t \in [t_0, +\infty).
\end{aligned} \qquad (3.96)$$

It means that they are close to the solution $z_0(t - t_0)$ lying on the separatrix which connects the saddle points z^* and z^{**} for the values of t belonging to the negative and positive semi-axis respectively.

Let $d(t_0)$ denote the orthogonal projection of the difference of solutions (3.96) at the moment $t = t_0$ into the line normal to the separatrix at the point $z_0(0)$. We get

$$d(t_0) = \varepsilon \frac{F\big(z_0(0)\big) \wedge \Big(z_1^-(t_0, t_0) - z_1^+(t_0, t_0)\Big)}{\big|F\big(z_0(0)\big)\big|} + O\left(\varepsilon^2\right). \qquad (3.97)$$

The function

$$M(t_0) = F\big(z_0(0)\big) \wedge \Big(z_1^-(t_0, t_0) - z_1^+(t_0, t_0)\Big) \qquad (3.98)$$

is called Melnikov's function. It is a certain measure of the relative distance between two points $z_\varepsilon^-(t_0, t_0)$ and $z_\varepsilon^+(t_0, t_0)$ which lie on the intersection of the plane $t = t_0$ and the surfaces S_-^* and S_+^{**} respectively. If the function $M(t_0)$ changes the sign as t_0 changes, then the points $z_\varepsilon^-(t_0, t_0)$, $z_\varepsilon^+(t_0, t_0)$ change their orientation with respect to the vector which is normal to the separatrix at the point $z_0(0)$. It means that for a sufficiently small ε the surfaces S_-^* and S_+^{**} intersect. Thus, the invariant lines L_-^* and L_+^{**} also intersect. If $M(t_0)$ does not change the sign, then S_-^* and S_+^{**} do not have common points for small values ε.

$2°$ Melnikov's function can be expressed by the right-hand side of the equation (3.91). Let us denote

$$\Delta^-(t, t_0) = F\big(z_0(t - t_0)\big) \wedge z_1^-(t, t_0), \qquad (3.99)$$

$$\Delta^+(t, t_0) = F\big(z_0(t - t_0)\big) \wedge z_1^+(t, t_0). \qquad (3.100)$$

It is evident that $M(t_0) = \Delta^-(t_0, t_0) - \Delta^+(t_0, t_0)$.

After differentiation of the equality (3.99) we get

$$\frac{d\Delta^-(t, t_0)}{dt} = F_z'\big(z_0(t-t_0)\big)\frac{dz_0(t-t_0)}{dt} \wedge z_1^-(t, t_0) + F\big(z_0(t-t_0)\big) \wedge \frac{dz_1^-(t, t_0)}{dt},$$

where F_z' is the Jacobian matrix (the derivative of the mapping $F : R^2 \to R^2$). For a fixed t_0 the functions $z_0(t - t_0)$ and $z_1^-(t, t_0)$ fulfil the differential equations

$$\frac{dz_0(t - t_0)}{dt} = F\big(z_0(t - t_0)\big), \qquad (3.101)$$

$$\frac{dz_1^-(t, t_0)}{dt} = F_z'\big(z_0(t - t_0)\big)z_1^-(t, t_0) + f\big(z_0(t - t_0), t\big). \qquad (3.102)$$

From the above it follows that

$$\frac{d\Delta^-}{dt} = F_z'(z_0)\, F(z_0) \wedge z_1^- + F(z_0) \wedge F_z'(z_0)\, z_1^- + F(z_0) \wedge f(z_0, t) \equiv$$

$$\equiv \operatorname{div} F(z_0)\Delta^- + F(z_0) \wedge f(z_0, t),$$

and finally, and more precisely, we get

$$\frac{d\Delta^-(t,t_0)}{dt} = \operatorname{div} F\big(z_0(t-t_0)\big)\Delta^-(t,t_0) + F\big(z_0(t-t_0)\big) \wedge f\big(z_0(t-t_0),t\big).$$
(3.103)

Similarly we can see that the function $\Delta^+(t,t_0)$ fulfils exactly the same linear differential equation.

3° The solution of the equation (3.103) takes the form

$$\Delta^-(t,t_0) = \frac{\Delta^-(t_0,t_0)}{G(t-t_0)} + \int_{t_0}^t \frac{G(\tau-t_0)}{G(t-t_0)} F\big(z_0(\tau-t_0)\big) \wedge f\big(z_0(\tau-t_0),\tau\big)d\tau,$$
(3.104)

where the function $G(t)$ defined by (3.94) is the solution of the homogeneous differential equation

$$\frac{dG(t)}{dt} = -\operatorname{div} F\big(z_0(t)\big)G(t), \qquad G(0) = 1.$$

From (3.104) it follows that

$$\Delta^-(t_0,t_0) = -\int_{t_0}^t G(\tau-t_0)\,F\big(z_0(\tau-t_0)\big)\wedge f\big(z_0(\tau-t_0),\tau\big)d\tau + \Delta^-(t,t_0)G(t-t_0).$$
(3.105)

We come now to the proof that

$$\lim_{(t-t_0)\to-\infty} \Delta^-(t,t_0)G(t-t_0) = 0. \tag{3.106}$$

Let $\lambda_-^* < 0$ and $\lambda_+^* > 0$ denote two eigenvalues of the Jacobian matrix $F_z'(z^*)$ at the saddle point z^*. If $t \to -\infty$ then $z_0(t) = z^* + C(t)e^{\lambda_+^* t}$, where $C(t)$ is a bounded function. Since $z_1^-(t,t_0)$ is a bounded function for $t < t_0$, then there exists a positive number B_1 such that

$$|\Delta^-(t,t_0)| \le B_1 \exp\big(\lambda_+^*(t-t_0)\big) \quad \text{for } t < t_0,$$

which follows from (3.99) and from the asymptotic equality $F\big(z_0(t)\big) \approx$ $\approx F_z'(z^*)C(t)\,e^{\lambda_+^* t}$. Similarly, from (3.94) and from the equality $\operatorname{div} F(z^*) =$ $= \lambda_-^* + \lambda_+^*$ it follows that there exists a positive number B_2 such that

$$|G(t-t_0)| \le B_2 \exp\big(-(\lambda_-^* + \lambda_+^*)(t-t_0)\big) \quad \text{for } t < t_0.$$

Finally (3.106) holds.

Combining (3.105) and (3.106) we get

$$\Delta^-(t_0,t_0) = \int_{-\infty}^{t_0} G(\tau-t_0)\,F\big(z_0(\tau-t_0)\big) \wedge f\big(z_0(\tau-t_0),\tau\big)d\tau. \tag{3.107}$$

Similarly, we can see that

$$-\Delta^+(t_0, t_0) = \int_{t_0}^{\infty} G(\tau - t_0) \, F\big(z_0(\tau - t_0)\big) \wedge f\big(z_0(\tau - t_0), \tau\big) d\tau. \quad (3.108)$$

The sum of two components (3.107) and (3.108) yields the analytic form of Melnikov's function

$$M(t_0) = \int_{-\infty}^{+\infty} G(\tau - t_0) \, F\big(z_0(\tau - t_0)\big) \wedge f\big(z_0(\tau - t_0), \tau\big) d\tau \quad (3.109)$$

or the equivalent representation (3.95).

It finishes the proof of theorem. $\square$

The above theorem (with the same Melnikov's function) relates also to the homoclinic trajectory in the case when $z^* = z^{**}$ is the same saddle point.

3.7   Boundaries of attractive domains

We are interested in the attractive domain of a stable periodic solution of equation (3.4) which lies in a neighborhood of the stable singular point of averaged system (3.5). Boundary of this attractive domain is determined by invariant lines L_+ of the Poincaré mapping which are attracted to the hyperbolic fixed point of this mapping. For the system (3.4) with a small high-frequency perturbation $\varepsilon w(\varphi, \omega t)$ the invariant lines L_+ and L_- attracted to and repelled from the hyperbolic fixed point lie in a neighborhood of separatrices of a saddle point of averaged system as long as they are near this point. If the distance from the saddle point is not small then the lines L_+ and L_- can be either smooth curves or extremely complicated curves (see Fig.3.18), if a homoclinic trajectory exists. In this last case we observe fractal boundary of the attractive domain.

In this section we use Mielnikov's theorem to establish some sufficient conditions of the existence of homoclinic trajectories of the system

$$\frac{d\varphi}{d\tau_w} = v,$$

$$\frac{dv}{d\tau_w} = -\sin\varphi + \Delta - \frac{v}{\sqrt{a}} + \varepsilon_0 \sin(\varphi + c\tau_w). \quad (3.110)$$

This system is equivalent to (3.4), where

$$\tau_w = \frac{\sqrt{a}}{\nu}\tau, \quad v = \sqrt{a}(x + \Delta), \quad c = \frac{2\nu}{\sqrt{a}}, \quad \varepsilon_0 \sin(\varphi + c\tau_w) = \varepsilon w(\varphi, \tau).$$

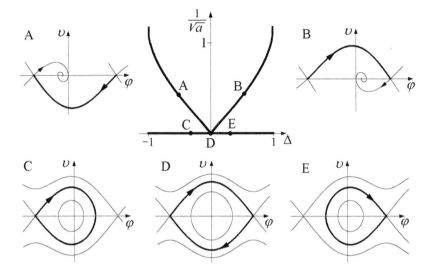

Fig. 3.21   The set of values of parameters Δ, $a^{-1/2}$ for which the system (3.30) has trajectories from a saddle point to a saddle point and graphs of these separatrices.

The equation (3.110) describes also a pendulum with an additional periodical perturbation.

For $\varepsilon_0 = 0$ the autonomous system depends on parameters Δ and $a^{-1/2}$. A separatrix from one saddle point to the other exists for the parameters lying on the lines shown in Fig.3.21. This follows from the results presented in Sections 3.2.4 and 3.2.7. Five different types of separatrices are shown in the figure. We are interested in such values of parameters for which there exist homoclinic or heteroclinic trajectories of the Poincaré mapping in neighborhoods of these separatrices.

3.7.1   Small values of the parameters: Δ, $a^{-1/2}$, ε_0

The system (3.110) is a particular case of (3.91), where

$$t = \tau_w, \quad z = (\varphi, v), \quad F_1(z) = v, \quad F_2(z) = -\sin\varphi,$$

$$\varepsilon = 1, \quad f_1(z,t) = 0, \quad f_2(z,t) = \Delta - \frac{v}{\sqrt{a}} + \varepsilon_0 \sin(\varphi + c\tau_w).$$

We assume that the positive parameters Δ, $a^{-1/2}$ and ε_0 are small.

The family of phase-plane trajectories of the autonomous conservative system

$$\frac{d\varphi}{d\tau_w} = v, \qquad \frac{dv}{d\tau_w} = -\sin\varphi \qquad (3.111)$$

takes the form $v^2 - 2\cos\varphi = E$, where E is a constant value. For $E = 2$ we have two separatrices

$$v = v_{\text{sep}}(\varphi) \quad \text{and} \quad v = -v_{\text{sep}}(\varphi), \quad \text{where } v_{\text{sep}}(\varphi) = 2\cos(\varphi/2), \quad (3.112)$$

connecting the saddle points $(-\pi, 0)$ and $(\pi, 0)$ (see Fig.3.21D).

The solution of the equations (3.111) lying on the separatrix satisfies the differential equation $\dfrac{d\varphi_s}{d\tau_w} = \pm 2\cos(\varphi_s/2)$, and it takes the form

$$\varphi_s(\tau_w) = \pm 2\arctan\left(\sinh\tau_w\right), \qquad v_s(\tau_w) = \pm\frac{2}{\cosh\tau_w}, \qquad (3.113)$$

where the upper and lower signs relate to the upper and lower separatrix respectively. The origin of the time scale $\tau_w = 0$ is chosen such that $\varphi_s(\tau_w)$ is an odd function and $v_s(\tau_w)$ is an even function.

The Melnikov function

$$M(t_0) = \Delta \int_{-\infty}^{+\infty} v_s(t)dt - \frac{1}{\sqrt{a}} \int_{-\infty}^{+\infty} v_s^2(t)dt + \varepsilon_0 \int_{-\infty}^{+\infty} v_s(t)\sin\left(\varphi_s(t) + ct + ct_0\right)dt$$

$$(3.114)$$

is the sum of three integrals which can be effectively calculated. For this we use the equations (3.111)–(3.113) and the formula of integrating by parts, if necessary.

$$\int_{-\infty}^{+\infty} v_s(t)dt = \pm \int_{-\pi}^{+\pi} d\varphi = \pm 2\pi,$$

$$\int_{-\infty}^{+\infty} v_s^2(t)dt = \pm \int_{-\pi}^{+\pi} v_{\text{sep}}(\varphi)d\varphi = \int_{-\pi}^{+\pi} 2\cos(\varphi/2)d\varphi = 8,$$

$$\int_{-\infty}^{+\infty} v_s(t)\sin\left(\varphi_s(t) + ct\right)dt = 0,$$

$$\int_{-\infty}^{+\infty} v_s(t)\cos\left(\varphi_s(t) + ct\right)dt = -\frac{1}{2}\int_{-\infty}^{+\infty}\left(cv_s^2(t) - 2c^2v_s(t)\right)\cos(ct)dt =$$

$$= -2\int_{-\infty}^{+\infty}\left(\frac{c}{\cosh^2 t} \mp \frac{c^2}{\cosh t}\right)\cos(ct)dt = -\frac{4\pi c^2}{\sinh c\pi}\exp\left(\mp c\pi/2\right).$$

Finally we get

$$M(t_0) = \pm 2\pi\Delta - \frac{8}{\sqrt{a}} - 2\pi\varepsilon_0 P_\pm(c)\sin ct_0, \qquad (3.115)$$

where

$$P_{\pm}(c) = \frac{2c^2}{\sinh c\pi} \exp\left(\mp c\pi/2\right). \tag{3.116}$$

Graphs of the function $P_+(c)$ for the upper separatrix and $P_-(c)$ for the lower separatrix are shown in Fig.3.22.

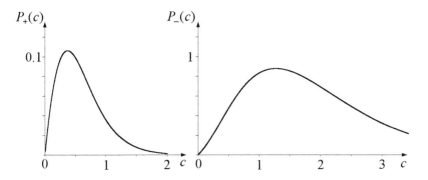

Fig. 3.22   Graphs of the functions (3.116).

If the parameters Δ, $a^{-1/2}$ and ε_0 are sufficiently small, and if

$$\left| \Delta \mp \frac{4}{\pi\sqrt{a}} \right| < \varepsilon_0 P_{\pm}(c), \tag{3.117}$$

then in a neighborhood of the upper or lower separatrix (3.112) there exist heteroclinic trajectories of the system (3.110).

The inequality (3.117) expressed by physical parameters ω, Ω, A, T of the PLL system takes the form

$$\left| \frac{\Omega - \omega}{A\Omega} \mp \frac{4}{\pi} \frac{1}{\sqrt{A\Omega T}} \right| < \varepsilon_0 P_{\pm}\left(\frac{2\omega T}{\sqrt{A\Omega T}}\right). \tag{3.118}$$

If $\varepsilon_0 = 0$ then $M(t_0) = \pm 2\pi\Delta - 8a^{-1/2}$ is a constant function, and the equality $M(t_0) = 0$ gives the asymptotic relation between small parameters Δ and $a^{-1/2}$ for which the autonomous system (3.5) has a separatrix from one saddle point to the other (compare (3.24)).

3.7.2   Large values of a

The system (3.110) is a particular case of (3.91), where

$$t = \tau_w, \quad z = (\varphi, v), \quad F_1(z) = v, \quad F_2(z) = -\sin\varphi + \Delta,$$

$$\varepsilon = 1, \quad f_1(z,t) = 0, \quad f_2(z,t) = -\frac{v}{\sqrt{a}} + \varepsilon_0 \sin(\varphi + c\tau_w).$$

We assume that the parameters $a^{-1/2}$ and ε_0 are small.

The family of phase trajectories of the non-perturbed conservative system

$$\frac{d\varphi}{d\tau_w} = v, \quad \frac{dv}{d\tau_w} = -\sin\varphi + \Delta \tag{3.119}$$

takes the form $v^2 - 2\cos\varphi - 2\Delta\varphi = E$. For $|\Delta| < 1$ there exists the saddle point $(\varphi_0, 0)$, where $\varphi_0 = \pi - \arcsin\Delta$, and the separatrix

$$v_s^2 - 2(\cos\varphi_s - \cos\varphi_0) - 2\Delta(\varphi_s - \varphi_0) = 0 \tag{3.120}$$

as shown in Fig.3.21C,E.

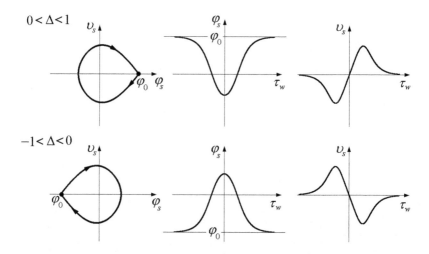

Fig. 3.23   Graphs of separatrices and solutions of Eq. (3.119) lying on these separatrices for $\Delta \in (0,1)$ and for $\Delta \in (-1,0)$.

The solution $\varphi_s(\tau_w)$, $v_s(\tau_w)$ of the system (3.119) lying on this separatrix can not be expressed by elementary functions but it can be determined numerically. Graphs of the solution are shown in Fig.3.23. The origin of the time scale $\tau_w = 0$ is chosen such that $\varphi_s(\tau_w)$ is an even function and $v_s(\tau_w)$ is an odd function.

Melnikov's function takes the form

$$M(t_0) = \frac{1}{\sqrt{a}} \int_{-\infty}^{+\infty} v_s^2(t)\,dt + \varepsilon_0 \int_{-\infty}^{+\infty} v_s(t) \sin\big(\varphi_s(t) + ct + ct_0\big)dt =$$

$$= \frac{1}{\sqrt{a}} S(\Delta) + \varepsilon_0 |\Gamma(\Delta, c)| \sin\big(ct_0 + \arg\Gamma(\Delta, c)\big), \tag{3.121}$$

where

$$S(\Delta) = \int_{-\infty}^{+\infty} v_s^2(t)dt = \int_{-\infty}^{+\infty} v_s(t)\frac{d\varphi_s(t)}{dt}dt$$

is the surface area enveloped by the separatrix loop, and

$$\Gamma(\Delta, c) = \int_{-\infty}^{+\infty} v_s(t)\exp\left(j\varphi_s(t) + jct\right)dt.$$

If $|\Delta|$ increases from 0 to 1, then the area $S(\Delta)$ decreases from 16 to zero.

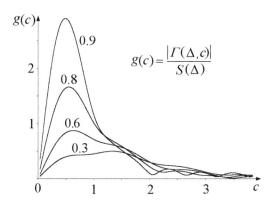

Fig. 3.24 Coefficient from the formula (3.122) versus normalized frequency c for $\Delta =$ $= 0.3, 0.6, 0.8, 0.9$.

If the parameters $a^{-1/2}$ and ε_0 are sufficiently small and if

$$\varepsilon_0\frac{|\Gamma(\Delta, c)|}{S(\Delta)} > \frac{1}{\sqrt{a}}, \qquad (3.122)$$

then, in a neighborhood of the separatrix (3.120), there exist homoclinic trajectories of the system (3.110). The numerically determined coefficient $|\Gamma(\Delta, c)|/S(\Delta)$ is shown in Fig.3.24 as a function of c for four values of $|\Delta|$.

3.7.3 A neighborhood of the line $|\Delta| = H(a)$

Let us denote by a^*, Δ^* such values of parameters for which the autonomous system

$$\frac{d\varphi}{d\tau_w} = v, \qquad \frac{dv}{d\tau_w} = -\sin\varphi + \Delta^* - \frac{v}{\sqrt{a^*}} \qquad (3.123)$$

has a separatrix from the saddle point $(\varphi_0 - 2\pi, 0)$ to the saddle point $(\varphi_0, 0)$, where $\varphi_0 = \pi - \arcsin\Delta^*$, as shown in Fig.3.21A,B. The equality

$|\Delta^*| = H(a^*)$ holds (compare with Sec. 3.2.4). The system (3.110) is a particular case of (3.91), where

$$t = \tau_w, \quad z = (\varphi, v), \quad F_1(z) = v, \quad F_2(z) = -\sin\varphi + \Delta^* - \frac{v}{\sqrt{a^*}},$$

$$\varepsilon = 1, \quad f_1(z, t) = 0, \quad f_2(z, t) = \varepsilon_\Delta - \varepsilon_a v + \varepsilon_0 \sin(\varphi + c\tau_w),$$

$$\varepsilon_\Delta = \Delta - \Delta^*, \quad \varepsilon_a = \frac{1}{\sqrt{a}} - \frac{1}{\sqrt{a^*}}.$$

We assume that the parameters ε_Δ, ε_a and ε_0 are small.

Let $\varphi_s(\tau_w)$, $v_s(\tau_w)$ denote the solution of the system (3.123) lying on the separatrix. Melnikov's function takes the form

$$
\begin{aligned}
M(t_0) &= \int_{-\infty}^{+\infty} v_s(t)\left(\varepsilon_\Delta - \varepsilon_a v_s(t) + \varepsilon_0 \sin\left(\varphi_s(t) + ct + ct_0\right)\right) \exp\left(\frac{t}{\sqrt{a^*}}\right) dt = \\
&= A\varepsilon_\Delta - B\varepsilon_a + \varepsilon_0 |\Gamma| \sin\left(ct_0 + \arg\Gamma\right),
\end{aligned}
\tag{3.124}
$$

where the numbers A, B, Γ are expressed by the convergent integrals

$$A = \int_{-\infty}^{+\infty} v_s(t) \exp\left(\frac{t}{\sqrt{a^*}}\right) dt,$$

$$B = \int_{-\infty}^{+\infty} \left(v_s(t)\right)^2 \exp\left(\frac{t}{\sqrt{a^*}}\right) dt,$$

$$\Gamma = \int_{-\infty}^{+\infty} v_s(t) \exp\left(\frac{t}{\sqrt{a^*}} + j\varphi_s(t) + jct\right) dt.$$

If the parameters ε_Δ, ε_a and ε_0 are sufficiently small and if

$$|A\varepsilon_\Delta - B\varepsilon_a| < \varepsilon_0 |\Gamma|,
\tag{3.125}$$

then in a neighborhood of the separatrix of the system (3.123) there exist heteroclinic trajectories of the system (3.110). The coefficients A, B and $|\Gamma|$ depend on a^* (or equivalently on Δ^*) and, moreover, $|\Gamma|$ depends on c.

The coefficients $A/|\Gamma|$ and $B/|\Gamma|$ versus c, for several values of Δ^*, are shown in Fig.3.25

3.7.4 *Numerical experiments*

In the plane of parameters $\Delta, a^{-1/2}$ the domains of existence of homoclinic trajectories, given by (3.117), (3.122), (3.125) for small values of ε_0, usually are more extensive for greater values of ε_0, e.g. for $\varepsilon_0 = 1$. Examples of

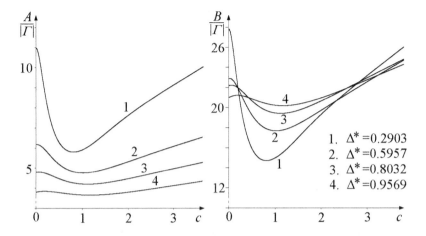

Fig. 3.25   Coefficients from the formula (3.125) versus normalized frequency c for four values of the parameter Δ^*.

the invariant lines L_+^*, L_-^*, L_+^{**}, L_-^{**} of the hyperbolic fixed points z^*, z^{**} of the Poincaré mapping of the equation (3.36) are shown in Fig.3.26.

Intersection points of lines L_+, L_- form homoclinic (or heteroclinic) trajectories. On a cylinder $S \times R$ we have $z^* \equiv z^{**}$ and heteroclinic trajectories in the plane are homoclinic on the cylinder.

Four types of such trajectories are possible:

$L_-^{**} \cap L_+^*$ shown in Fig.3.26b,c,d,

$L_-^* \cap L_+^*$ shown in Fig.3.26c,

$L_-^{**} \cap L_+^{**}$ shown in Fig.3.26d,

$L_-^* \cap L_+^{**}$ does not shown in the figure (it exists only in a very small range of parameters if $\Delta > 0$).

Let s denote the stable fixed point of the Poincaré mapping i.e. the value $s = z_{sp}(0)$ of the stable periodic solution of the equation (3.36) at the point $\tau = 0$. If there is no homoclinic trajectory, then the attractive domain of this point is bounded by smooth invariant lines L_+^{**} and L_+^* attracted to hyperbolic fixed points as in Fig.3.26a.

If a homoclinic trajectory exists then (probably) the same lines form the boundary of the attractive domain but, in this case, it has an extremely complicated shape. An example is shown in Fig.3.27. For the same values of parameters as shown in Fig.3.26d, the following numerical experiment is given.

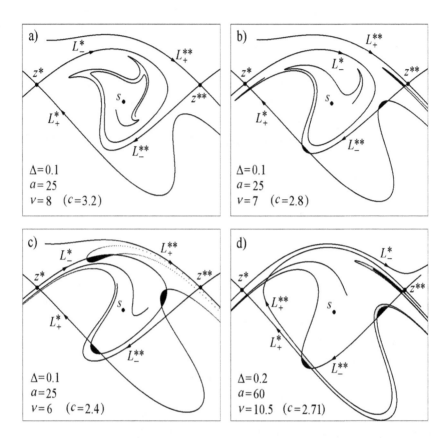

Fig. 3.26   Examples of invariant lines of hyperbolic fixed points of the Poincaré mapping of Eq. (3.36). Range of variables: $|\varphi| < 3.8$, $x \in (-1; 0.4)$ for pictures (a), (b), (c) and $|\varphi| < 3.8$, $x \in (-0.7; 0.1)$ for picture (d).

For each point z, from the set of 760×760 points of the rectangle $\{\varphi, x : |\varphi| < 3, \ x \in (-0.7; 0.1)\}$, the segment of trajectory $\{P^k(z)\}$ of the Poincaré mapping is calculated for $k = 1, ..., 50$. If the trajectory does not leave the rectangle then the point z is marked either as a black point, if the trajectory tends to the point s, or as a grey point, if we do not know what happens with the next segments of the trajectory. If we take a longer segment of the trajectory (e.g. $k = 1, ..., 100$), then some grey points disappear (the trajectory leaves the rectangle to the left or to the right), some change into black points (the trajectory tends to s), and some remain in the rectangle even if $k \to \infty$. However, the measure of these grey points usually tends to zero if k increases. This phenomenon is noted as so called transient chaos.

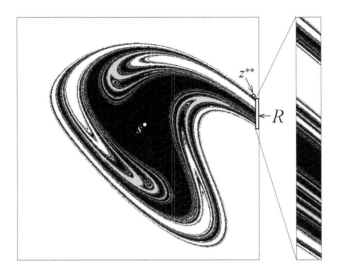

Fig. 3.27 The attractive domain (black) of the stable fixed point s of the Poincaré mapping. The hyperbolic fixed point z^{**} is also marked. The rectangle R is enlarged 8-times and shown on the right-hand side of the figure.

In the next section it will be proved that in a neighborhood of a homoclinic trajectory there exists an infinite number of unstable periodic trajectories and an uncountable set of non-periodic (chaotic) trajectories which remain in a bounded domain and do not tend to any stable orbit.

3.8 The Smale horseshoe. Transient chaos

This section is recommended only to the readers interested in trajectories which start from very small neighborhood of homoclinic trajectory (from a domain of very small measure).

Let there exists a homoclinic trajectory of the point z^*. Let us consider the quadrangle Q with vertices $ABCD$ situated near the point z^* as shown in Fig.3.28. Side AB lies on the invariant line L_-. Sides BC and AD are approximately parallel to L_+ and lie on the right of L_+. The Poincaré mapping contracts the quadrangle Q in the direction of the line L_+ and stretches it in the direction of L_-. If the rectangle Q lies in a sufficiently small neighborhood of the point z^* then the rates of contraction and stretching are determined by the eigenvalues of the Jacobian matrix of the Poincaré mapping at the fixed point z^*. Successive iterations of the Poincaré mapping P transform the set Q onto the sets $P(Q), P^2(Q), \ldots, P^n(Q)$ which are in a thin stretched shape. All these sets lie on the right of L_-. Thus, there exists such a number n for which $P^n(Q)$ is in the shape of a "horseshoe" which crosses the quadrangle Q along two strips. Such a mapping $F = P^n$ is called the *Smale horseshoe* in Q. Evidently, there exists infinite integers k and quadrangles Q_k such that for everyone Q_k the mapping P^k is a Smale's horseshoe in Q_k.

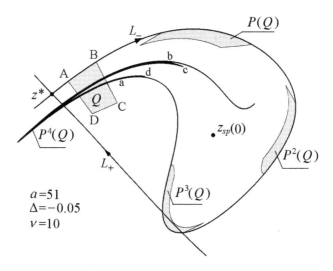

Fig. 3.28   Segments of the invariant lines L_+, L_- of the hyperbolic fixed point $z^* =$ $= P(z^*)$ of the Poincaré mapping of Eq. (3.36) and the sets $P^k(Q)$, $k = 1, 2, 3, 4$ for a given quadrangle Q. The mapping $F = P^4$ is the Smale horseshoe in Q.

A simple example of such a mapping was given by Stephen Smale in 1961. This example is important for the theory of dynamical systems (see Refs. [20], [56]). Below we give a geometrical description of the Smale horseshoe and prove that there exists an infinite number of periodic points of the mapping F in the set $Q \cap F(Q)$.

3.8.1   Invariant set of the Smale horseshoe

Let $Q \subset R^2$ be a square of side length equal to one. Let us define the map $F : R^2 \to R^2$ as a composition of two maps $F = F_2 \circ F_1$. The map F_1 contracts the square in a horizontal direction and stretches it in a vertical direction. Suppose that the set $F_1(Q)$ is a rectangle of side lengths $\lambda < \frac{1}{2}$

and $\mu > 2$. The map F_2 bends this rectangle onto the horseshoe $F(Q)$ which crosses Q along two vertical strips V_0 and V_1 of width λ (Fig.3.29). The inverse map $F^{-1} = F_1^{-1} \circ F_2^{-1}$ also transforms the square Q onto the horseshoe $F^{-1}(Q)$ which crosses the square Q along two horizontal disjoint strips H_0 and H_1 of width μ^{-1}. The equalities $Q \cap F(Q) = V_0 \cup V_1$ and $F^{-1}(Q) \cap Q = H_0 \cup H_1$ hold. If we restrict the domain of F to the strip H_0 then the map is linear and $F(H_0) = V_0$. Similarly, in the strip H_1 the map F is linear and $F(H_1) = V_1$. The set $F^{-1}(Q) \cap Q \cap F(Q)$ consists of four rectangles, the horizontal sides of which are of length λ and the vertical sides are μ^{-1}. Fig.3.30 shows how the sets $F^2(Q)$ and $F^{-2}(Q)$ look.

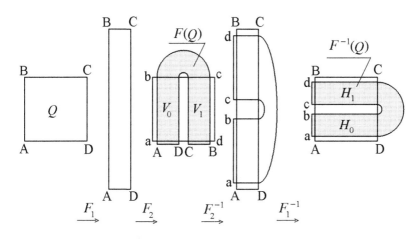

Fig. 3.29   The Smale horseshoe.

It is easily seen that the sets

$$Q \cap F(Q) \cap \ldots \cap F^n(Q) \quad \text{and} \quad F^{-m}(Q) \cap \ldots \cap F^{-1}(Q) \cap Q$$

consist of 2^n vertical strips of width λ^n and of 2^m horizontal strips of width μ^{-m} respectively. The set $\bigcap_{i=-m}^{n} F^i(Q)$ consists of 2^{n+m} rectangles, the horizontal and vertical sides of which are λ^n and μ^{-m} respectively. The area of the union of these rectangles is equal to $(2\lambda)^n (\frac{1}{2}\mu)^{-m}$ and tends to zero as $n, m \to \infty$.

To each of such rectangles we give an address described as a number in the binary system. The address is a sequence consisting of zeros and ones, and containing $n + m$ terms. Each term has its place denoted by the index changing from $-n$ to $m - 1$.

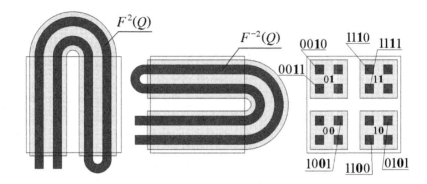

Fig. 3.30   The sets $F^2(Q)$, $F^{-2}(Q)$, $F^{-2}(Q) \cap F^2(Q)$ (in black) and the sets $F^1(Q)$, $F^{-1}(Q)$, $F^{-1}(Q) \cap F^1(Q)$ (in grey) for a given square Q and the Smale horseshoe F.

The address is defined as follows. The sequence

$$a = \{a_{-n}, \ldots, a_{-1}, a_0, a_1, \ldots, a_{m-1}\} \qquad (3.126)$$

is called the *address* of a rectangle $Q_{n,m}^a \subset \bigcap_{i=1-m}^n F^i(Q)$ if

$$a_i = \begin{cases} 0 & \text{when } F^i(Q_{n,m}^a) \subset H_0, \\ 1 & \text{when } F^i(Q_{n,m}^a) \subset H_1. \end{cases} \qquad (3.127)$$

for each $i = -n, \ldots, 0, \ldots, m-1$. Addresses of several rectangles are shown in Fig.3.30 (the digit of index $i = 0$ is marked in bold).

If we adjoin a digit a_{-n-1} at the beginning (or a_m at the end) of address (3.126) of the rectangle $Q_{n,m}^a$, we obtain addresses of two rectangles: $Q_{n+1,m}^{0a}$ for $a_{-n-1} = 0$ and $Q_{n+1,m}^{1a}$ for $a_{-n-1} = 1$ (or $Q_{n,m+1}^{a0}$ for $a_m = 0$ and $Q_{n,m+1}^{a1}$ for $a_m = 1$ respectively). All these rectangles are contained in $Q_{n,m}^a$.

Rectangle $Q_{n,m}^a$ which has the address (3.126) contains all rectangles with addresses $\{\ldots, b_{-n-1}, a_{-n}, \ldots, a_0, \ldots, a_{m-1}, c_m, \ldots\}$, where $\{b_i\}$ for $i \le -n-1$ and $\{c_i\}$ for $i \ge m$ are arbitrary finite sequences of zeros and ones.

From the above it follows that the map F takes some points out of the square Q and puts some other points in it from the outside. Moreover, in Q there exists the invariant set

$$\Lambda = \bigcap_{i=-\infty}^{+\infty} F^i(Q) = \{z : F^i(z) \in Q, \ -\infty < i < +\infty\} \qquad (3.128)$$

of points which stay in Q for all iterations of the maps F and F^{-1}. Each point of the set Λ is an intersection point of a vertical line belonging to the set $\bigcap_{i=0}^{\infty} F^i(Q)$ and a horizontal line belonging to the set $\bigcap_{i=-\infty}^{0} F^i(Q)$.

The measure (area) of the set $\Lambda = F(\Lambda)$ is equal to zero. The set Λ has the cardinal number of the continuum and is nowhere-dense.

3.8.2 Homeomorphism

Let S denote the set of all doubly infinite sequences (the index takes on all integer values)

$$a = \{a_i\}, \quad b = \{b_i\}, \quad c = \{c_i\}, \dots \tag{3.129}$$

the terms of which are zeroes and ones. The distance between two sequences a and b is defined by the formula

$$\rho(a,b) = \sum_{i=-\infty}^{+\infty} 2^{-|i|} |a_i - b_i|. \tag{3.130}$$

It is a well defined metric in the space S.

In the set S we define an operation σ which shifts all terms of a sequence by one place to the left:

$$\text{if } a = \{a_i\}, \quad \text{then } \sigma(a) = \{b_i\}, \quad \text{where } b_i = a_{i+1}. \tag{3.131}$$

Theorem 3.8. *There exists such a homeomorphism h (one-to-one correspondence which is continuous in both directions) of the set Λ onto the set S, that for each $z \in \Lambda$*

$$\text{if } h(z) = a \quad \text{then } h\big(F(z)\big) = \sigma(a). \tag{3.132}$$

Proof.   The map $h : \Lambda \to S$, which associates addresses to points of the set Λ, is defined as follows:

$$\text{if } h(z) = \{a_i\}, \quad \text{then } a_i = \begin{cases} 0 & \text{when } F^i(z) \in H_0, \\ 1 & \text{when } F^i(z) \in H_1. \end{cases} \tag{3.133}$$

From (3.133) we have (3.132), because $F^i\big(F(z)\big) = F^{i+1}(z)$. The map h is defined for all $z \in \Lambda$, which follows from $\Lambda = \bigcap_{i=-\infty}^{+\infty} F^i(Q) \subset F^{-1}(Q) \cap Q = H_0 \cup H_1$.

If $z_a \neq z_b$ then for a sufficiently large N the points z_a and z_b belong to two disjoint rectangles $Q_{N,N}^a$, $Q_{N,N}^b$ and, consequently, they have different addresses. Hence $h(z_a) \neq h(z_b)$, and h is injective. The inverse map h^{-1} is defined on the whole set S, i.e. every sequence $a = \{a_i\}$ is the image of a point $z_a \in \Lambda$. Indeed, each finite sequence $\{a_{-N}, \ldots, a_0, \ldots, a_N\}$ corresponds to exactly one closed rectangle $Q_{N,N}$, and $Q_{N,N} \supset Q_{N+1,N+1}$. The point $z_a = h^{-1}(a)$ is a limit of a sequence of nested closed rectangles with diameters tending to zero as $N \to \infty$.

The continuity of the maps h and h^{-1} follows also from the properties of addresses of rectangles. If two points z_a and z_b of the set Λ are sufficiently close, i.e. if $|z_a - z_b| < \varepsilon$, then one can find such a number $N(\varepsilon)$ (where $N \to \infty$ as $\varepsilon \to 0$) that z_a and z_b belong to a common rectangle $Q_{N,N}$. Then the sequences $a = h(z_a)$ and $b = h(z_b)$ fulfil the condition $a_i = b_i$ for $|i| \leq N$ and, consequently, $\rho(a, b) \leq 2 \sum_{i=N+1}^{\infty} 2^{-i} = 2^{-N+1} \to 0$ as $\varepsilon \to 0$. Conversely, if two sequences a and b from the set S are sufficiently close, i.e. if $\rho(a, b) < \varepsilon$, then the equality $a_i = b_i$ is satisfied for $|i| \leq N(\varepsilon)$. Thus, the points $z_a = h^{-1}(a)$ and $z_b = h^{-1}(b)$ of the set Λ belong to the same rectangle $Q_{N,N}$ of a small diameter.   □

From (3.132) it follows that $F(z) = h^{-1} \sigma h(z)$ and, consequently,

$$F^n(z) = h^{-1} \sigma^n h(z) \quad \text{for } z \in \Lambda, \tag{3.134}$$

or, equivalently,

$$\sigma^n(a) = h F^n h^{-1}(a) \quad \text{for } a \in S \tag{3.135}$$

for each integer n.

Each topological property of iterations of the map F in the set Λ has an equivalent in the set of properties of the operation σ defined on S. The next theorem is a direct consequence of this fact.

Theorem 3.9. *The invariant set Λ of the map F has the following properties:*

(a) *it contains a countable set of periodic orbits of periods equal to all natural numbers,*

(b) *it contains an uncountable set of non-periodic orbits,*

(c) *it contains a dense trajectory,*

(d) *every map $\tilde{F}$ sufficiently C^1-close to the map F has an invariant set $\tilde{\Lambda}$ which has the same properties* (a),(b),(c). *The map $\tilde{F}$ defined on $\tilde{\Lambda}$ is topologically equivalent to the map F defined on Λ.*

Proof.   (a) In the set S there is a countable number of periodic sequences (for each period n there is at most 2^n sequences). For each n-periodic sequence $a = \{a_i\}$ of the set S, the point $z_a = h^{-1}(a)$ of the set Λ is a periodic point of the map F. Indeed, from $\sigma^n(a) = a$ it follows that $F^n(z_a) = z_a$, according to (3.134).

(b) The set S contains an uncountable set of non-periodic sequences. If a is non-periodic, then the trajectory $\{F^n(z_a)\}$ of the point $z_a = h^{-1}(a)$ is non-periodic in the set Λ.

(c) Let us form the sequence $a = \{a_i\}$ in the following way. For $i < 0$ we take a_i absolutely arbitrarily. For $i \geq 0$ we take successively all 4 two-element subsequences $00, 01, 10, 11$, then all 8 three-element subsequences $000, 010, 100, 110, 001, 011, 101, 111$, and all 16 four-element subsequences etc. From (3.130) it follows that for any $b \in S$ and for any $\varepsilon > 0$ there exists such a natural number $n = n(\varepsilon)$ that $\rho\big(b, \sigma^n(a)\big) < \varepsilon$. Indeed, two sequences b and $\sigma^n(a)$ are close to each other if for a sufficiently large number $N(\varepsilon)$ all terms with indexes $|i| < N(\varepsilon)$ are identical. Thus, the set $\sigma^k(a)$ for $k = 1, 2, \ldots$ is dense in S and, consequently, the semi-trajectory $\{F^k(z_a)\}$, $k = 1, 2, \ldots$, where $z_a = h^{-1}(a)$ is dense in Λ.

(d) Intuitively, this property follows immediately from the description of the map F. The precise proof is omitted. $\qquad\qquad\square$

It is convenient to extend the notion of the Smale horseshoe. Every map $F : Q \to R^2$ which has the invariant set $\Lambda \subset Q$ with the properties given by Theorem 3.9 we call the Smale horseshoe.

Using Theorem 3.8 we give below additional properties of the Smale horseshoe the proofs of which are left to the readers.

Proposition 3.2. *The invariant set Λ of the Smale horseshoe has the following properties:*

(e) *For any two periodic trajectories $\{F^n(z_a)\}$, $\{F^n(z_b)\}$ lying in Λ there exists such a point $z_c \in \Lambda$, that its positive semi-trajectory tends to $\{F^n(z_a)\}$ and its negative semi-trajectory tends to $\{F^n(z_b)\}$.*

(f) *for a periodic trajectory $\{F^n(z_a)\}$ lying in Λ there exists such a point $z_b \in \Lambda$ that the positive (negative) semi-trajectory of this point tends to $\{F^n(z_a)\}$ and negative (positive) semi-trajectory is dense in Λ.*

3.8.3   Comments

Let us return to the origin of the Smale horseshoe presented at the beginning of this section. Let z^* be a hyperbolic fixed point of the Poincaré mapping of the system (3.82). It can be proved that if the invariant lines attracted to and repelled from the point z^* intersect transversally (if there exists a homoclinic trajectory), then in a neighborhood of the point z^* there exist quadrangles Q such that some iteration P^n of the Poincaré mapping becomes the Smale horseshoe in Q. Equation (3.82) with initial conditions lying in Q has a family of "chaotic" solutions, i.e. an infinite number of un-stable periodic solutions of long periods and an uncountable set of bounded non-periodic solutions. In the set Q there are no stable trajectories.

From numerical experiments it follows that, after sufficiently long time, all numerically determined trajectories exit Q and, usually, tend to a stable limit set. Usually the numerically determined initial value does not belong to the invariant set Λ, because Λ is a nowhere-dense set and mes $\Lambda = 0$. Thus one observes the existence of the Smale horseshoe in numerical ex-periments as an extremely long-time transient state called *transient chaos*.

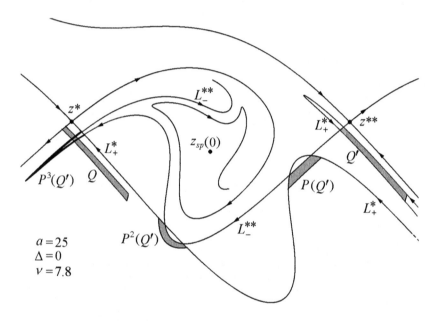

Fig. 3.31   An example of the Smale horseshoe $F(\varphi, x) = P^3(\varphi + 2\pi, x)$ for Eq. (3.36).

Example 3.2. The Poincaré mapping of the equation (3.36) is investi-gated for $a = 25$, $\Delta = 0$, $\nu = 7.8$. Fig.3.31 shows numerically determined invariant lines attracted to and repelled from the hyperbolic fixed points $z^* = (\varphi^*, x^*)$ and $z^{**} = (\varphi^* + 2\pi, x^*)$. If $\varphi \bmod 2\pi$ is treated as a cyclic variable, then z^* and z^{**} are the same point on the cylinder $S \times R$, and the invariant lines L_-^{**}, L_+^* intersect along a homoclinic trajectory on the cylinder (but heteroclinic in φx-plane).

Let us take the quadrangle Q' as shown in Fig.3.31. One side of Q' lies on L_{-}^{**} and two sides lie on L_{+}^{*}. The set $P^3(Q')$ looks like a horseshoe which crosses the quadrangle $Q = Q' - (2\pi, 0)$. Quadrangles Q and Q' overlap on the cylinder $S \times R$. The mapping $F(\varphi, v) = P^3(\varphi + 2\pi, v)$ is the Smale horseshoe with an invariant set $\Lambda \subset F(Q) \cap Q$. Consequently, for an initial point from the set Λ the system shown in Fig.3.1 has unstable periodic or chaotic output signal $\cos\theta(t)$ which phase θ increases (approximately) by 4π during the time $6\pi/\omega$.

3.9 Higher order systems reducible to the second order ones

3.9.1 *The system with a filter of the higher order*

Let us consider the PLL system shown in Fig.3.32. Let the transfer function $H(s)$ of the low-pass filter be a rational function analytical for $\text{Re } s \geq 0$, and $H(0) = 1$, $H(\infty) = 0$. For simplicity, we assume that the function $H(s)$ has simple poles only. So, the function $H(s)$ can be represented by

$$H(s) = \sum_{i=0}^{N} \frac{p_i \alpha_i}{s + \alpha_i}, \quad \text{where } \text{Re } \alpha_i > 0, \quad i = 0, 1, ..., N. \quad (3.136)$$

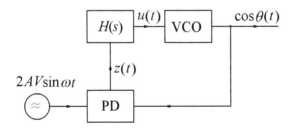

Fig. 3.32   A higher order phase-locked loop.

The parameters α_i, p_i are either real or complex numbers. If α_i, p_i are complex, then in the sum (3.136) there exists the pair of conjugate complex terms

$$\frac{p_i \alpha_i}{s + \alpha_i} + \frac{\overline{p_i \alpha_i}}{s + \overline{\alpha_i}} = \frac{2\text{Re}(p_i\alpha_i)s + 2\text{Re}(p_i)|\alpha_i|^2}{s^2 + 2\text{Re}(\alpha_i)s + |\alpha_i|^2},$$

because the rational transfer function $H(s)$ has real coefficients.

An alternative description of the filter uses differential equations. Let $z(t)$ be an input, and $u(t)$ be an output signal of the filter. Then

$$u(t) = \sum_{i=0}^{N} u_i(t), \quad \text{where} \quad \frac{du_i(t)}{dt} + \alpha_i u_i(t) = p_i \alpha_i z(t). \qquad (3.137)$$

We assume that p_0 and α_0 are real and the first term in (3.136) is the dominant one (the parameters $|p_i|$ for $i = 1, 2, ..., N$ are small). If we eliminate u_0 from (3.137) replacing it by $u - (u_1 + ... + u_N)$, we obtain

$$\frac{du(t)}{dt} + \alpha_0 u(t) = \alpha_* z(t) + \sum_{i=1}^{N} (\alpha_0 - \alpha_i) u_i(t),$$

where

$$\alpha_* = \sum_{i=0}^{N} p_i \alpha_i = \lim_{s \to \infty} sH(s). \qquad (3.138)$$

The coefficient α_* is a weighted mean value of coordinates α_i of singular points with weights p_i, where $\sum_{i=0}^{N} p_i = H(0) = 1$. Moreover, α_* is the value of *impulse response* of the filter at the moment $t = 0$.

Let the equations of the voltage controlled oscillator (VCO) and phase detector (PD) be the same as in Section 3.1:

$$\frac{d\theta(t)}{dt} = \Omega \left(1 + \frac{u(t)}{V}\right) \quad \text{and} \quad z(t) = 2AV \sin \omega t \cos \theta(t).$$

So, the equations of the PLL system take the form

$$\frac{d\theta}{dt} = \Omega \left(1 + \frac{u}{V}\right), \qquad (3.139)$$

$$\frac{du}{dt} = -\alpha_0 u + 2\alpha_* AV \sin \omega t \cos \theta + \sum_{i=1}^{N} (\alpha_0 - \alpha_i) u_i, \qquad (3.140)$$

$$\frac{du_i}{dt} = -\alpha_i u_i + 2p_i \alpha_i AV \sin \omega t \cos \theta, \quad i = 1, 2, ..., N. \qquad (3.141)$$

In the next subsections two problems will be discussed: the existence of a stable two-dimensional integral manifold and the transfer function of the PLL system linearized in a neighborhood of a stable fixed point of autonomous equations.

3.9.2 Two-dimensional integral manifold

If there exists a globally stable two-dimensional integral manifold then the dynamics of the system (3.139)–(3.141) is the same as the dynamics of a two-dimensional system.

If $p_i = 0$ for $i = 1, 2, ..., N$, then the equations (3.141) have the well known solutions

$$u_i(t) = u_i(t_0)e^{-\alpha_i(t-t_0)}. \tag{3.142}$$

In the set of functions bounded for $t \in (-\infty, +\infty)$ the functions $u_i(t) \equiv 0$ form the unique solution, and this solution is globally stable. Geometrically speaking, the system of equations (3.139)–(3.141), for $p_i = 0$, $i = 1, 2, ..., N$, has the globally stable two-dimensional integral manifold

$$\{(\theta, u, u_1, ..., u_N, t) : u_i = 0 \text{ for } i = 1, 2, ..., N, (\theta, u) \in R^2, t \in R\} \tag{3.143}$$

embedded in $(2 + N + 1)$-dimensional space.

We will show that if

$$\alpha_0 + \sqrt{2A\Omega\alpha_*} < \min_{1 \leq i \leq N} \operatorname{Re} \alpha_i \tag{3.144}$$

and if the numbers $|p_i|$ for $i = 1, 2, ..., N$ are sufficiently small, then the system of equations (3.139)–(3.141) has a globally stable integral manifold. More precisely, we have the following theorem:

Theorem 3.10. *Let us denote*

$$g(t) = \sum_{i=1}^{N} p_i\alpha_i(\alpha_0 - \alpha_i)e^{-\alpha_i t}. \tag{3.145}$$

If there exists a positive number σ such that

$$\int_0^{\infty} |g(t)|e^{\sigma t}dt \leq \frac{2\alpha_*}{3}\left(\frac{\sigma - \alpha_0}{\sqrt{2A\Omega\alpha_*}} - 1\right), \tag{3.146}$$

then in a neighborhood of the manifold (3.143) the system of equations (3.139)–(3.141) has a globally stable integral manifold

$$\{(\theta, u, u_1, ..., u_N, t) : u_i = h_i(t, \theta, u), i = 1, 2, ..., N, (\theta, u, t) \in R^3\}, \tag{3.147}$$

where the Lipschitzian functions $h_i(t, \theta, u)$ are bounded by numbers directly proportional to $|p_i|$. The functions $h_i(t, \theta, u)$ are 2π-periodic with respect to ωt and θ.

For the initial conditions lying on the integral manifold, the system (3.139)–(3.141) is described by equations

$$\frac{d\theta}{dt} = \Omega\left(1 + \frac{u}{V}\right), \tag{3.148}$$

$$\frac{du}{dt} = -\alpha_0 u + 2\alpha_* AV \sin \omega t \cos \theta + H(t, \theta, u), \tag{3.149}$$

where

$$H(t,\theta,u) = \sum_{i=1}^{N}(\alpha_0 - \alpha_i)h_i(t,\theta,u). \qquad (3.150)$$

The proof of the theorem is given in the next subsection. In Fig.3.33 there is shown the scheme of integral manifold (3.147), a family of solutions of equations (3.148)–(3.149) lying on the manifold and a graph of one solution of equations (3.139)–(3.141) attracted to the manifold.

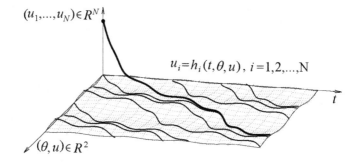

Fig. 3.33   Graphs of solutions of Eqs. (3.139)–(3.141) lying on the integral manifold (3.147) and a graph of a solution attracted to the manifold (bold line).

Remark 1. If the number σ satisfies the condition (3.146) then it belongs to the open interval the end-points of which are given by the left--hand side and right-hand side of the inequality (3.144). If σ lies outside of this interval, then either the right-hand side of (3.146) is negative or the left-hand side of (3.146) is equal to $+\infty$.

Remark 2. Although the functions $h_i(t,\theta,u)$ can be complex-valued, the function $H(t,\theta,u)$ is always real-valued.

Example 3.3. Let us consider the PLL system with the filter which has the following transfer function:

$$H(s) = \frac{0.09}{sT + 0.1} + \frac{0.2}{sT + 2}, \quad \text{where } T = \frac{1}{\Omega}$$

(Ω is the quiescent frequency of VCO). The parameters α_i, p_i take the values:

$$\alpha_0 = 0.1\,\Omega, \quad \alpha_1 = 2\,\Omega, \quad p_0 = 0.9, \quad p_1 = 0.1, \quad \alpha_* = 0.29\,\Omega$$

and, consequently, $g(t) = -0.38\,\Omega^2 e^{-2\Omega t}$.

The inequality (3.146) takes the form

$$\frac{0.38\,\Omega^2}{2\Omega - \sigma} \leq 0.29\,\Omega\,\frac{2}{3}\left(\frac{\sigma - 0.1\,\Omega}{\sqrt{0.58\,A\Omega^2}} - 1\right)$$

and it has a positive solution σ if and only if $A \leq 0.17460$. For these values of A there exists a globally stable two-dimensional integral manifold, and the dynamics of the 3-dimensional PLL system is the same as that of a two-dimensional one.

3.9.3   *Proof of Theorem 3.10*

The idea of the proof is as in Theorem 3.3. We only give the basic steps of the proof. The details (e.g. proofs of two lemmas) are left to the reader.

The proof will be divided into three steps:

1° estimation of solutions of equations (3.148)–(3.149) for a given $H(t, \theta, u)$,

2° construction of an equation of the integral manifold,

3° application of the contracting mapping principle to prove that the equation has exactly one solution.

The proof of the stability of the integral manifold will be omitted.

1° Let us consider the system of equations (3.148)–(3.149) for a given bounded continuous function $H(t, \theta, u)$ which satisfies the Lipschitz condition

$$|H(t, \theta_1, u_1) - H(t, \theta_2, u_2)| \leq \lambda_a|\theta_1 - \theta_2| + \lambda_b|u_1 - u_2| \qquad (3.151)$$

for all values $(t, \theta, u) \in R^3$.

Let

$$\theta = \theta_H(t; t^0, \theta^0, u^0), \qquad u = u_H(t; t^0, \theta^0, u^0) \qquad (3.152)$$

denote the solution of equations (3.148)–(3.149) which starts from an initial point (t^0, θ^0, u^0), i.e. such a solution that

$$\theta_H(t^0; t^0, \theta^0, u^0) = \theta^0, \qquad u_H(t^0; t^0, \theta^0, u^0) = u^0.$$

Let us introduce the following notations

$$c = \sqrt{\frac{A\Omega}{2\alpha_*}}, \qquad \sigma = \alpha_0 + 2c\alpha_* + \lambda_b + \frac{c\lambda_a}{2}$$

and

$$\|H\| = \sup_{R^3} |H(t, \theta, u)|. \qquad (3.153)$$

The following lemma gives some estimation of the difference of the two solutions of equations (3.148)–(3.149) respectively for two different functions H_1, H_2 and two different initial points (t^0, θ_1^0, u_1^0), (t^0, θ_2^0, u_2^0).

Lemma 3.4. *Let* $\theta_{H1}(t; t^0, \theta_1^0, u_1^0)$ *and* $\theta_{H2}(t; t^0, \theta_2^0, u_2^0)$ *be the solutions of equations* (3.148)–(3.149) *for given continuous functions* $H_1(t, \theta, u)$, $H_2(t, \theta, u)$ *and initial points* (t^0, θ_1^0, u_1^0), (t^0, θ_2^0, u_2^0) *respectively.*

If the functions H_1 and H_2 are bounded and if they satisfy the Lipschitz condition (3.151), *then*

$$|\theta_{H1}(t; t^0, \theta_1^0, u_1^0) - \theta_{H2}(t; t^0, \theta_2^0, u_2^0)| \leq$$

$$\leq \frac{c}{\sigma}\left(e^{\sigma|t - t^0|} - 1\right)||H_1 - H_2|| + \left(|\theta_1^0 - \theta_2^0| + c|u_1^0 - u_2^0|\right)e^{\sigma|t - t^0|} \qquad (3.154)$$

for $t \in (-\infty, +\infty)$.

The proof of Lemma 3.4 is omitted. It is almost the same as the proof of the Lemma 3.1 in Section 3.4.3. Similar estimation can be obtained for $|u_{H1} - u_{H2}|$, but it is not interesting for us.

2° Now, an equation of the integral manifold will be constructed. Let the function $\theta_H(t; t^0, \theta^0, u^0)$ be the solution of equations (3.148)–(3.149) for a given function H. Substituting this solution for θ in (3.141) we obtain N linear differential equations of the first order

$$\frac{du_i(t)}{dt} + \alpha_i u_i(t) = 2AV p_i \alpha_i \sin \omega t \cos \theta_H(t; t^0, \theta^0, u^0) \qquad (3.155)$$

for $i = 1, 2, ..., N$. In the set of functions bounded for all $t \in (-\infty, +\infty)$, each of the equations (3.155) has exactly one solution

$$u_i(t; t^0, \theta^0, u^0) = 2AV p_i \alpha_i \int_0^\infty e^{-\alpha_i \tau} \sin \omega(t - \tau) \cos \theta_H(t - \tau; t^0, \theta^0, u^0) d\tau \qquad (3.156)$$

which depends on the initial point (t^0, θ^0, u^0) and on the function H.

Now let us choose the function H as follows:

$$H(t, \theta, u) = \sum_{i=1}^{N} (\alpha_0 - \alpha_i) u_i(t; t, \theta, u). \qquad (3.157)$$

This function satisfies the equation

$$H(t, \theta, u) = 2AV \int_0^\infty g(\tau) \sin \omega(t - \tau) \cos \theta_H(t - \tau; t, \theta, u) d\tau \qquad (3.158)$$

which we call the *equation of the integral manifold*. If there exists a solution $H_*(t, \theta, u)$ of the equation (3.158), then the function $\theta_{H_*}(t; t^0, \theta^0, u^0)$

satisfies equations (3.148)–(3.149) with $H = H_*$. Consequently, the set of the functions

$$h_i(t, \theta, u) = 2AV p_i \alpha_i \int_0^\infty e^{-\alpha_i \tau} \sin \omega(t - \tau) \cos \theta_{H_*}(t - \tau; t, \theta, u) d\tau \quad (3.159)$$

for $i = 1, 2, ..., N$ determines the integral manifold (3.147).

In the next step of the proof we show that the equation (3.158) has exactly one solution.

$3°$ Let $\mathcal{H}_\lambda$ denote the set of continuous real-valued functions $H(t, \theta, u)$ defined on R^3, satisfying the Lipschitz condition (3.151) with a constant $\lambda = \{\lambda_a, \lambda_b\}$ and bounded by the constant $2 \int_0^\infty |g(t)| dt$. The set $\mathcal{H}_\lambda$ with the metric

$$\rho(H_1, H_2) = ||H_1 - H_2|| = \sup_{R^3} |H_1(t, \theta, u) - H_2(t, \theta, u)| \quad (3.160)$$

is a complete metric space. Let G denote the operator defined on the space $\mathcal{H}_\lambda$ by the right-hand side of the equation (3.158).

Lemma 3.5. *If a positive number σ satisfies the inequality (3.146) then, for the parameter $\lambda = \{\lambda_a, \lambda_b\}$ given by*

$$\lambda_b = c\lambda_a = \frac{2}{3}\left(\sigma - \alpha_0 - \sqrt{2A\Omega\alpha_*}\right), \quad (3.161)$$

the operator G maps the space $\mathcal{H}_\lambda$ into itself and satisfies the Lipschitz condition $||G(H_1) - G(H_2)|| \le p||H_1 - H_2||$ with the constant

$$p = \frac{\sqrt{2A\Omega\alpha_*}}{\alpha_* \sigma} \int_0^\infty |g(t)|(e^{\sigma t} - 1) dt < 1. \quad (3.162)$$

The proof of Lemma 3.5 is omitted. It follows almost immediately from the formulae (3.158) and (3.154).

Now we can finish the proof of Theorem 3.10. By the contraction mapping principle (see Ref. [24]), in the space $\mathcal{H}_\lambda$ with the parameter $\lambda = \{\lambda_a, \lambda_b\}$ given by (3.161), the equation (3.158) has exactly one solution. Consequently, the integral manifold (3.147) exists.

3.9.4   *The local linearization*

Let us consider the PLL system (Fig.3.34b) with the input modulated wave

$$2AV(1 + a(t)) \sin(\omega t + \alpha(t)). \quad (3.163)$$

The output signal is denoted by $\cos\left(\omega t + \varphi(t)\right)$. Let the equation of the phase detector PD take the form

$$z(t) = -AV\left(1 + a(t)\right)\sin\left(\varphi(t) - \alpha(t)\right). \tag{3.164}$$

The system is described by the equations

$$\frac{d\varphi(t)}{dt} = (\Omega - \omega) + \Omega\frac{u(t)}{V}, \qquad u(t) = \sum_{i=0}^{N} u_i(t),$$

$$\frac{du_i(t)}{dt} + \alpha_i u_i(t) = \alpha_i p_i z(t), \tag{3.165}$$

where $z(t)$ is given by (3.164).

If $a(t) \equiv 0$ and $\alpha(t) \equiv 0$, as shown in Fig.3.34 a, and if $|\Delta| < 1$ (where $\Delta = \frac{\Omega - \omega}{A\Omega}$), then the system (3.164)–(3.165) is autonomous and has equilibrium points (the constant solutions)

$$\varphi = \varphi_*, \quad \text{where} \quad \sin\varphi_* = \Delta, \quad u_{i*} = -p_i AV\Delta, \quad z_* = u_* = -AV\Delta. \tag{3.166}$$

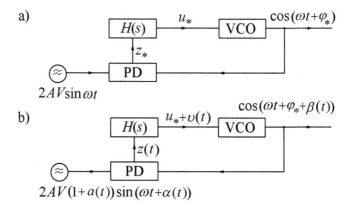

Fig. 3.34   The PLL system with two different input signals: a sine-wave and a modulated wave.

Now, let $a(t)$ and $\alpha(t)$ take small values:

$$|a(t)| \ll 1, \qquad |\alpha(t)| \ll \pi,$$

and let new coordinates be introduced in a neighborhood of the equilibrium point

$$\beta(t) = \varphi(t) - \varphi_*, \qquad v_i(t) = u_i(t) + p_i AV\Delta, \qquad v(t) = u(t) + AV\Delta.$$

The nonlinear system (3.164)–(3.165) linearized in a neighborhood of the point (3.166) takes the form

$$\frac{d\beta(t)}{dt} = \Omega\frac{v(t)}{V}, \qquad v(t) = \sum_{i=0}^{N} v_i(t),$$

$$\frac{dv_i(t)}{dt} + \alpha_i v_i(t) = -\alpha_i p_i AV\Big(a(t)\sin\varphi_* + \big(\beta(t) - \alpha(t)\big)\cos\varphi_*\Big). \tag{3.167}$$

In technical literature [9], [28] the system (3.167) is usually investigated using the method of the Laplace transformation.

Assume that $a(t) \equiv 0$ and $\alpha(t) \equiv 0$ for $t \leq 0$ and denote the Laplace transforms of functions $\beta(t)$, $v_i(t)$, $a(t)$, $\alpha(t)$ by $\tilde{\beta}(s)$, $\tilde{v}_i(s)$, $\tilde{a}(s)$, $\tilde{\alpha}(s)$ respectively.

The differential equations (3.167) with the zero initial conditions are equivalent to algebraic equations for the Laplace transforms

$$s\tilde{\beta}(s) = \Omega\frac{\tilde{v}(s)}{V}, \qquad \tilde{v}(s) = \sum_{i=0}^{N} \tilde{v}_i(s),$$

$$\big(s + \alpha_i\big)\tilde{v}_i(s) = -\alpha_i p_i AV\Big(\tilde{a}(s)\sin\varphi_* + \big(\tilde{\beta}(s) - \tilde{\alpha}(s)\big)\cos\varphi_*\Big). \tag{3.168}$$

Extracting $\tilde{\beta}(s)$ from (3.168) we get

$$\tilde{\beta}(s) = \frac{A\Omega H(s)\Big(\tilde{\alpha}(s)\cos\varphi_* - \tilde{a}(s)\sin\varphi_*\Big)}{s + A\Omega H(s)\cos\varphi_*}. \tag{3.169}$$

The PLL systems are used as amplifiers of phase-modulated signals or as amplitude-modulated-to-phase-modulated converters.

In the first case we have $a(t) \equiv 0$ and the transfer function of amplifier is

$$\frac{\tilde{\beta}(s)}{\tilde{\alpha}(s)} = \frac{A\Omega H(s)\cos\varphi_*}{s + A\Omega H(s)\cos\varphi_*}. \tag{3.170}$$

It happens that the input signal (3.163) with the time-varying phase $\alpha(t)$ has an amplitude perturbed by a noise signal $a(t)$. If $\omega = \Omega$ then $\varphi_* = 0$ and the amplifier of phase-modulated signals is not sensitive to small perturbation of the amplitude of input signal.

In the second case $\alpha(t) \equiv 0$, and the transfer function of converter is

$$\frac{\tilde{\beta}(s)}{\tilde{a}(s)} = -\frac{A\Omega H(s)\sin\varphi_*}{s + A\Omega H(s)\cos\varphi_*}. \tag{3.171}$$

The system (3.164)–(3.165) is locally stable in a neighborhood of the equilibrium point (3.166) if and only if all poles of the rational function

$$R(s) = \frac{H(s)}{s + A\Omega H(s)\cos\varphi_*} \qquad (3.172)$$

lie in the half plane $\operatorname{Re} s < 0$. Using formulae (3.136),(3.138) we can calculate coefficients of polynomials in the numerator and denominator of the function

$$R(s) = \frac{\alpha_* s^N + \ldots + r}{s^{N+2} + \ldots + rA\Omega\cos\varphi_*}, \qquad \text{where} \quad r = \prod_{i=0}^{N}\alpha_i > 0. \qquad (3.173)$$

If $\cos\varphi_* < 0$ then the equilibrium point (3.166) is unstable because the first and the last coefficients of denominator have opposite signs. If $\cos\varphi_* > 0$ then the equilibrium point can be stable or unstable. It depends on the transfer function $H(s)$ of the filter and on the amplitude A of the input signal.

Example 3.4. Let the transfer function of the second-order filter take the form $H(s) = \left(T^2 s^2 + 2bTs + 1\right)^{-1}$, where b, T are positive parameters. According to the well known Hurwitz's theorem the rational function

$$R(s) = \frac{1}{T^2 s^3 + 2bT s^2 + s + A\Omega\cos\varphi_*}$$

is analytical for $\operatorname{Re} s \geq 0$ if and only if

$$0 < TA\Omega\cos\varphi_* < 2b.$$

These inequalities are necessary and sufficient for the local stability of the equilibrium point.

Chapter 4

One-dimensional discrete-time Phase-Locked Loop

4.1 Recurrence equations of the system

Let us consider the system shown in Fig.4.1. Let $U_{\text{out}}\big(\theta(t)\big)$ be an output signal of the voltage controlled oscillator (VCO). The shape of the graph of a 2π-periodic function $U_{\text{out}}(\theta)$ is not essential for us.

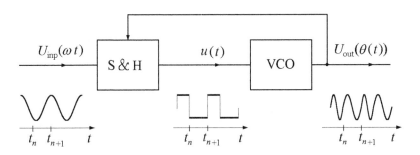

Fig. 4.1   The discrete-time phase-locked loop system.

The instantaneous frequency $\frac{d\theta}{dt}$ of the output signal depends on the controlling voltage $u(t)$ in the following way:

$$\frac{d\theta}{dt} = \Omega\left(1 + \frac{u(t)}{V}\right) \tag{4.1}$$

where Ω is the quiescent frequency of VCO and V is the value of controlling voltage which doubles the generator frequency.

At the moment t_n, when

$$\theta(t_n) = 2\pi n, \qquad n = 0, 1, 2, \ldots, \tag{4.2}$$

the input signal $U_{\text{inp}}(\omega t)$ is sampled in a *sample-and-hold* unit (S&H), and after that the value of controlling voltage remains constant

$$u(t) = U_{\text{inp}}(\omega t_n) \quad \text{for} \quad t \in (t_n, t_{n+1}). \tag{4.3}$$

The equality

$$\theta(t_{n+1}) - \theta(t_n) = \Omega \left(1 + \frac{U_{\text{inp}}(\omega t_n)}{V} \right) (t_{n+1} - t_n) \tag{4.4}$$

follows immediately from (4.1). Its left-hand side is equal to 2π. Hence,

$$\text{for} \quad \tau_n = \omega t_n \quad \text{we have} \quad \tau_{n+1} = T(\tau_n), \tag{4.5}$$

where

$$T(\tau) = \tau + \frac{\omega}{\Omega} \cdot \frac{2\pi}{1 + U_{\text{inp}}(\tau)/V}. \tag{4.6}$$

The variable $\tau = \omega t$ denotes a dimensionless time or a phase of the input signal at the moment t.

For example, if $U_{\text{inp}}(\tau) = AV \sin \tau$, then

$$T(\tau) = \tau + \frac{\omega}{\Omega} \cdot \frac{2\pi}{1 + A \sin \tau}. \tag{4.7}$$

Another case is also possible (see Ref. [46]). Let the unit S&H produce at the moment t_n a puls of duration shorter than $\min_i(t_{i+1}-t_i)$ with an area $kU_{\text{inp}}(\omega t_n)$ proportional to the value of the input signal at the moment t_n. Then, from (4.1) we get

$$2\pi = \theta(t_{n+1}) - \theta(t_n) = \Omega(t_{n+1} - t_n) + k\Omega \frac{U_{\text{inp}}(\omega t_n)}{V}$$

and, finally,

$$t_{n+1} = t_n + \frac{2\pi}{\Omega} - \frac{k}{V} U_{\text{inp}}(\omega t_n).$$

The relation between the successive sampling moments is also given by (4.5) but the map T has a different form now. If $U_{\text{inp}}(\tau) = AV \sin \tau$, then

$$T(\tau) = \tau + 2\pi\mu + a \sin \tau, \quad \text{where} \quad \mu = \frac{\omega}{\Omega}, \quad a = -kA\omega. \tag{4.8}$$

If the parameter A takes small values and $k = 2\pi/\Omega$, then the maps (4.7) and (4.8) are close to each other.

The output signal of the *discrete-time phase-locked loop* (DPLL) with a sampling unit described by (4.3) is of the form

$$U_{\text{out}}(\theta(t)) = U_{\text{out}} \left(2\pi \frac{\omega t - \tau_n}{T(\tau_n) - \tau_n} \right) \quad \text{for } \omega t \in (\tau_n, \tau_{n+1}), \quad n = 0, 1, 2, \ldots \tag{4.9}$$

So, the oscillations in the DPLL system depend on the properties of the sequence $\{\tau_0, \tau_1, \tau_2, \ldots\}$ defined by the recursion formula (4.5) for various initial points τ_0 and for various values of the parameters $\frac{\omega}{\Omega}, A$.

Description of a discrete-time phase-locked loop by the map (4.5) is similar to description of a continuous-time phase-locked loop (without filter) by the Poincaré mapping. It might be seem that the discrete-time loop is easier to analyze because one avoids the difficult stage of determination of the Poincaré mapping from a differential equation. However, generally, the map T is non-monotonic (contrary to the Poincaré mapping) and this fact substantially complicates the dynamics of the system. Only in such a range of parameters $\frac{\omega}{\Omega}, A$ for which (4.7) (or respectively (4.8)) is an increasing function, the dynamics of the discrete-time phase-locked loop is the same as the dynamics of a continuous-time loop described by the first order differential equation.

Let us pay attention to one more distinction with respect to Chapter 2. During the investigation of a continuous-time loop, the phases $\{\theta_0, \theta_1, \theta_2, \ldots\}$ of the output signal were determined at the moments when the phase of the input signal was equal to $2\pi n$. Here, the situation is opposite. We determine the phases $\{\tau_0, \tau_1, \tau_2, \ldots\}$ of the input signal at the moments when the phases of the output signal are equal to $2\pi n$. This difference is inessential from the mathematical point of view, but it changes the physical interpretation of the rotation number.

The main aim of this chapter is to show some frequency locking regions in the plane of parameters $\frac{\omega}{\Omega}, A$ and regions, where DPLL has stable output signals. Moreover, bifurcations of periodic signals on the borders of these regions will be discussed.

4.2 Periodic output signals

Generally, a discrete-time phase-locked loop has several periodic solutions of different periods for various initial points τ_0. The output signal of DPLL is a periodic function if and only if it starts from a periodic point. In this section the basic properties of periodic points will be given.

4.2.1 *Type of a periodic point*

Let T be a continuous map satisfying identity

$$T(\tau + 2\pi) = T(\tau) + 2\pi. \tag{4.10}$$

It is evident that also $T^m(\tau + 2\pi) = T^m(\tau) + 2\pi$ for each natural number m and the functions $P_m(\tau) = T^m(\tau) - \tau$ are 2π-periodic.

Definition 4.1. The number τ_0 which satisfies the equation

$$T^m(\tau_0) = \tau_0 + 2\pi n \qquad (4.11)$$

and the condition

$$T^k(\tau_0) - \tau_0 \neq 0 \bmod 2\pi, \quad \text{for} \quad k = 1, 2, \ldots, m-1, \qquad (4.12)$$

is said to be a periodic point of the type n/m (of the map T). A periodic point of the type $0/1$ is called a fixed point.

We call the number τ_0 a periodic point of the map T if it is a periodic point of the type n/m for certain natural numbers n and m.

Each periodic point has only one type. If n and m are coprime numbers, then (4.12) follows from (4.11).

Let us consider the *trajectory* of an arbitrary point τ_0 of the circle S, i.e. the infinite sequence

$$\tau_0, \tau_1, \tau_2, \ldots \quad (\bmod 2\pi), \quad \text{where} \quad \tau_{n+1} = T(\tau_n).$$

This sequence can be:
 – *periodic* if τ_0 is a periodic point,
 – *pre-periodic* if τ_0 is not periodic, but there exists $N > 0$ such that τ_N is a periodic point,
 – *non-periodic* if it does not have two identical terms.
If τ_0 is a periodic point of the type n/m, then the finite sequence

$$\{\tau_0, \tau_1, \tau_2, \ldots, \tau_{m-1}\} \quad (\bmod 2\pi), \quad \text{where} \quad \tau_{n+1} = T(\tau_n), \qquad (4.13)$$

is called the *periodic orbit* of the point τ_0. We do not distinguish orbits which differ by a circular permutation of terms. Every point of the periodic orbit is also a periodic point of the type n/m. In the periodic orbit there are no identical terms.

If τ_0 is a periodic point of the type n/m of the map (4.6), then the output signal $U_{\text{out}}(\theta_0(t))$ of DPLL fulfilling the initial condition $\theta_0(\tau_0/\omega) = 0$ is a periodic signal. Its period is equal to $2\pi n/\omega$, and during this time the phase $\theta_0(t)$ increases by $2\pi m$. We call $U_{\text{out}}(\theta_0(t))$ the signal of the type n/m. If we take any two periodic points τ_i, τ_j of the same orbit as the initial points, then the corresponding signals $U_{\text{out}}(\theta_i(t))$, $U_{\text{out}}(\theta_j(t))$ overlap if one of them is shifted along the time axis by an interval of length equal to $(\tau_i - \tau_j)/\omega$.

4.2.2 Basic properties of periodic points

Several theorems concerned with periodic points of the map T which satisfy the condition (4.10) will be given.

Theorem 4.1. *If a continuous map T has a periodic point of the type n/m, then the rational number $\frac{n}{m}$ fulfils the inequalities*

$$\frac{1}{2\pi} \min_\tau \left(T(\tau) - \tau \right) \le \frac{n}{m} \le \frac{1}{2\pi} \max_\tau \left(T(\tau) - \tau \right). \tag{4.14}$$

Proof. By (4.10) it is obvious that $T(\tau) = P(\tau) + \tau$, where P is a 2π-periodic function. Putting $T(\tau)$ in place of τ, we get $T^2(\tau) = P\big(T(\tau)\big) + P(\tau) + \tau$ and, similarly,

$$T^m(\tau) = P\big(T^{m-1}(\tau)\big) + P\big(T^{m-2}(\tau)\big) + \ldots + P\big(T(\tau)\big) + P(\tau) + \tau. \tag{4.15}$$

Therefore,

$$m \min_t P(t) \le T^m(\tau) - \tau \le m \max_t P(t). \tag{4.16}$$

A periodic point τ_0 of the type n/m satisfies the equality $T^m(\tau_0) - \tau_0 = 2\pi n$ and, consequently, (4.14) follows from (4.16). □

Corollary 4.1. *If the map (4.7) has a periodic point of the type n/m, then*

$$\left| \frac{\omega}{\Omega} - \frac{n}{m} \right| \le \frac{n}{m} A.$$

If the map (4.8) has a periodic point of the type n/m, then

$$\left| \frac{\omega}{\Omega} - \frac{n}{m} \right| \le \frac{|a|}{2\pi}.$$

For $m = 1$ (and only for $m = 1$) the inequalities given above are also sufficient for the existence of periodic points of the type $n/1$.

It is easy to give examples of maps which have periodic points of different types. For example, the map $T(\tau) = \tau + \pi(3 + \sin \tau)$ has the periodic point $\tau_1 = -\frac{\pi}{2}$ of the type $1/1$ and the periodic point $\tau_2 = \frac{\pi}{2}$ of the type $2/1$. In such cases the map has an infinite number of periodic points of different types.

Theorem 4.2. *If a continuous map T has a periodic point τ_1 of the type n_1/m_1 and a periodic point τ_2 of the type n_2/m_2 where $\frac{n_1}{m_1} < \frac{n_2}{m_2}$, then for any coprime natural numbers n_0, m_0 fulfilling the inequality*

$$\frac{n_1}{m_1} < \frac{n_0}{m_0} < \frac{n_2}{m_2} \tag{4.17}$$

there exists a periodic point of the type n_0/m_0 of the map T.

Proof.   First, we will show two properties of iterates of the map T fulfilling the condition (4.10).

(a) For fixed n and m we denote $G(\tau) = T^m(\tau) - 2\pi n$. For any natural number p we have the identity

$$G^p(\tau) = T^{pm}(\tau) - p\,2\pi n. \qquad (4.18)$$

Indeed, $G^2(\tau) = T^m\big(T^m(\tau) - 2\pi n\big) - 2\pi n = T^{2m}(\tau) - 2\,(2\pi n)$ and by induction we get (4.18).

(b) Let $G : R \to R$ be a continuous map. If there exists α such that $G^k(\alpha) = \alpha$ for a natural number k, then there exists τ_0 such that $G(\tau_0) = \tau_0$. Indeed, in the opposite case we would have the inequality $G(\tau) > \tau$ or $G(\tau) < \tau$ for all τ. Putting $G(\tau)$ in place of τ we get $G^2(\tau) > \tau$ or $G^2(\tau) < \tau$ and, similarly, $G^k(\tau) > \tau$ or $G^k(\tau) < \tau$. Therefore, this contradicts the existence of a fixed point of the map G^k.

Now, we may initiate the proof of Theorem 4.2.

The periodic point τ_1 is a fixed point of the map $G_1(\tau) = T^{m_1}(\tau) - 2\pi n_1$, and for $p = m_0 m_2$ we have

$$\tau_1 = G_1^p(\tau_1) = T^{m_0 m_1 m_2}(\tau_1) - 2\pi n_1 m_0 m_2. \qquad (4.19)$$

The periodic point τ_2 is a fixed point of the map $G_2(\tau) = T^{m_2}(\tau) - 2\pi n_2$, and for $q = m_0 m_1$ we have

$$\tau_2 = G_2^q(\tau_2) = T^{m_0 m_1 m_2}(\tau_2) - 2\pi n_2 m_0 m_1. \qquad (4.20)$$

From (4.19) and (4.20) it follows that the continuous 2π-periodic function

$$P(\tau) = T^{m_0 m_1 m_2}(\tau) - \tau \qquad (4.21)$$

takes all values from the interval $[2\pi n_1 m_0 m_2,\ 2\pi n_2 m_0 m_1]$ for $\tau \in [\tau_1, \tau_2]$. So, there exists a point α such that $P(\alpha) = 2\pi n_0 m_1 m_2$, i.e.

$$T^{k m_0}(\alpha) - 2\pi n_0 k = \alpha, \qquad \text{where}\ \ k = m_1 m_2.$$

By the above property (b), there exists a point τ_0 such that

$$T^{m_0}(\tau_0) - 2\pi n_0 = \tau_0. \qquad (4.22)$$

We have assumed that n_0, m_0 are coprime, so τ_0 is a periodic point of the type n_0/m_0. $\qquad\square$

Frequently, the map T has periodic points of the type pn/pm, where p, n, m are natural numbers and n, m are coprime. It is evident, that a periodic point of the type pn/pm of the map T is a periodic point of the type $0/p$ of the map $G(\tau) = T^m(\tau) - 2\pi n$ or, equivalently, the fixed point of the map G^p.

Theorem 4.3 (Sharkovsky). *Let the set of all natural numbers be ordered as follows:*

$$3 \prec 5 \prec 7 \prec 9 \prec \ldots \prec 2 \cdot 3 \prec 2 \cdot 5 \prec 2 \cdot 7 \prec \ldots \prec 2^2 \cdot 3 \prec 2^2 \cdot 5 \prec 2^2 \cdot 7 \prec \ldots$$

$$\ldots \prec 2^n \cdot 3 \prec 2^n \cdot 5 \prec 2^n \cdot 7 \prec \ldots \prec \ldots \prec 2^5 \prec 2^4 \prec 2^3 \prec 2^2 \prec 2 \prec 1.$$

In the above sequence, successive odd numbers appear first, then come odd numbers multiplied by 2, then odd numbers multiplied by $2^2, 2^3, 2^4, \ldots$ and, at the end, powers of number 2 in descending order.

If a number p appears sooner than a number k (if $p \prec k$) and if the continuous map T has a periodic point of the type pn/pm where n and m are coprime, then T also has a periodic point of the type kn/km.

The proof is omitted. It may be found in Refs. [54], [55], [12], where the theorem is formulated for a continuous map $G : R \to R$ and is not specified for the map T which satisfies the condition (4.10).

Corollary 4.2. *If T has a periodic point of the type $3n/3m$ where n and m are coprime, then for every natural number p there exists a periodic point of the type pn/pm of the map T.*

4.2.3 Li and Yorke Theorem

Let as introduce the notion of an *address* of a trajectory

$$\tau_0, \tau_1, \tau_2, \tau_3, \ldots \qquad \text{where} \quad \tau_{n+1} = G(\tau_n). \tag{4.23}$$

Let L and N be two closed intervals. Let G be such a continuous map of the union $L \cup N$ into the real line R, that

$$G(L) \supset N \qquad \text{and} \qquad G(N) \supset L \cup N. \tag{4.24}$$

Examples of such maps are shown in Fig.4.2, 4.3 and 4.4.

An infinite sequence $A = \{I_0, I_1, I_2, \ldots\}$ of intervals I_k we call the *address* (see Ref. [15]) if the following two conditions are satisfied:

1) for each k, either $I_k = L$ or $I_k = N$,
2) if $I_k = L$ then $I_{k+1} = N$.

Here are two examples of addresses:

$$N_0 N N L N L N L N L N N N_{10} L N L \, N N N L \, N L \, N_{20} N N L \, N N N N N L N_{30} \ldots$$
$$N_0 L N \, N L N \, N N N L N_{10} N N N N N L N N N N_{20} N N N L N N N N_{30} \ldots$$

In the first address $I_n = L$ if and only if n is a prime number greater than 2. In the second address $I_n = L$ if and only if n is the square of a natural

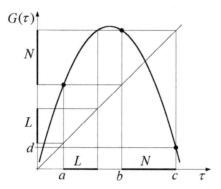

Fig. 4.2   An example of the map which satisfies the conditions (4.24).

number. For transparency of notation the terms with numbers n divisible by 10 are indicated.

Let $A^1 = \{I_0^1, I_1^1, I_2^1, ...\}$ and $A^2 = \{I_0^2, I_1^2, I_2^2, ...\}$ be two different addresses. We say that addresses A^1 and A^2 are *distant* if the intervals L, N are disjoint and if $I_n^1 \neq I_n^2$ for an infinite set of numbers n.

The set of all addresses has the cardinality of the continuum. The subset of addresses in which any two addresses are distant has the cardinality of the continuum also.

We say that $A(\tau_0) = \{I_0, I_1, I_2, ...\}$ is the *address of trajectory* (4.23) if $G^n(\tau_0) \in I_n$ for every $n = 0, 1, 2, ...$ If L and N are disjoint intervals, then each trajectory has no more then one address.

It will be proved that for every address one can find at least one trajectory (empty addresses do not exist).

Theorem 4.4 (about addresses). *If a map G is continuous in $L \cup N$ and satisfies conditions* (4.24), *then for every address $A = \{I_0, I_1, I_2, ...\}$ there exists $\tau_0 \in L \cup N$ such that $G^n(\tau_0) \in I_n$ for $n = 0, 1, 2, ...$.*

Proof.   At the beginning four properties of continuous functions will be described (see Ref. [32]).

Let I be a closed interval and let $F : I \to R$ be a continuous function.

(a) If I_1 is a closed interval and $I_1 \subset F(I)$, then there exists a closed interval $Q \subset I$ such that $F(Q) = I_1$.

(b) If $Q \subset I$ is a closed interval and if $F(Q) \supset Q$, then there exists a point $\tau_* \in Q$ such that $F(\tau_*) = \tau_*$.

(c) Let $\{I_n\}$, $n = 0, 1, 2, ...$ be a sequence of closed intervals such that

$G : I_n \to R$ is a continuous function and $I_{n+1} \subset G(I_n)$. Then there exists a sequence of closed intervals $\{Q_n\}$, $n = 0, 1, \ldots$ such that $Q_{n+1} \subset Q_n$ and $G^n(Q_n) = I_n$.

(d) Intersection of all intervals Q_n is not empty, and if $x \in \bigcap Q_k$, then $G^n(x) \in I_n$ for each $n = 0, 1, 2, \ldots$.

The properties (a) and (b) are evident, (c) will be proved by induction:

Let us assume that $Q_0 \equiv G^0(Q_0) = I_0$. Therefore, $I_1 \subset G(I_0) = G(Q_0)$, and the existence of a closed interval $Q_1 \subset Q_0$ such that $G(Q_1) = I_1$ follows from (a) for $F \equiv G$. So, we have proved (c) for $n = 1$. Suppose that (c) is true for a number $n-1$, i.e. there exists Q_{n-1} such that $G^{n-1}(Q_{n-1}) = I_{n-1}$ and $I_n \subset G(I_{n-1}) = G^n(Q_{n-1})$. By (a) for $F \equiv G^n$, there exists a closed interval $Q_n \subset Q_{n-1}$ such that $G^n(Q_n) = I_n$. Finally, (c) is true for all natural numbers n.

The property (d) is a simple consequence of (c).

By (4.24), each address satisfies assumptions of the property (c). So, by the property (d), there exists a point τ_0 such that $A = A(\tau_0)$ is the address of the trajectory $\{G^n(\tau_0)\}$. □

Example 4.1. Let us consider the map (4.8) with the parameters $\mu = 0.65$ and $a = -2.2$.

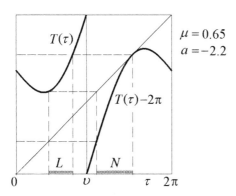

Fig. 4.3 Graph of the function (4.8) for selected values of parameters.

Graph of the function

$$G(\tau) = \begin{cases} T(\tau) & \text{for } \tau \in (0, v) \\ T(\tau) - 2\pi & \text{for } \tau \in (v, 2\pi) \end{cases} \quad \text{where } T(v) = 2\pi, \; v = 2.844\ldots$$

(or equivalently $G(\tau) = T(\tau) \bmod 2\pi$) is shown in Fig.4.3. There exist two closed intervals L and N which satisfy conditions (4.24). According to Theorem 4.4, for every address $A = \{I_0, I_1, I_2, \ldots\}$ one can find at least one trajectory $\{\tau_0, \tau_1, \tau_2, \ldots\}$ where $\tau_n = G(\tau_{n-1}) \in I_n$.

Let $\eta(n, A)$ denote the number of symbols N in the subset $\{I_0, I_1, \ldots, I_{n-1}\}$ of the address $A = \{I_0, I_1, I_2, \ldots\}$. The inequality $\frac{1}{2}(n-1) \leq$ $\leq \eta(n, A) \leq n$ holds. If A is the address of a trajectory $\{\tau_0, \tau_1, \tau_2, \ldots\}$ where $\tau_{n+1} = T(\tau_n)$, then $\tau_n - 2\pi\eta(n, A) \in L \cup N$. By Theorem 4.4, for every number $\rho_0 \in [0.5, 1]$ we can find an initial point τ_0 such that $\lim\limits_{n \to \infty} \dfrac{\tau_n}{2\pi n} = \rho_0$.

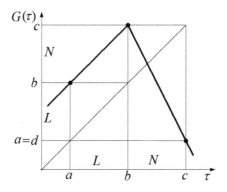

Fig. 4.4　An example of the function $G(\tau)$ which satisfies assumptions of Theorem 4.5.

Periodic trajectories of all periods can exist in the set of trajectories (4.23) (see Ref. [32]).

Theorem 4.5 (T.Y.Li, J.A.Yorke). *Let a continuous function G maps an interval I into itself. If there exists a point $a \in I$ such that the points $b = G(a)$, $c = G(b)$, $d = G(c)$ satisfy the inequalities*

$$d \leq a < b < c \qquad or \qquad d \geq a > b > c, \qquad (4.25)$$

then for every natural number p there exists $\tau^* \in I$ for which $G^p(\tau^*) = \tau^*$ and $G^k(\tau^*) \neq \tau^*$ for $k = 1, 2, 3, \ldots, p-1$.

Proof. Assume that $d \leq a < b < c$. In the case $d \geq a > b > c$ the proof goes similarly and we do not repeat it. Let us denote $L = [a, b]$, $N = [b, c]$. For a fixed integer $p > 1$ we choose the p-periodic sequence $\{I_n\}$ in the following way: $I_0 = I_1 = \ldots = I_{p-2} = N$, $I_{p-1} = L$ and next $I_{n+p} = I_n$

for $n = 0, 1, 2, \ldots$ If $p = 1$ then $I_n = N$ for all n. The conditions (4.24) are satisfied.

Let $\{Q_n\}$ be the sequence of closed intervals which was described in the proof of theorem 4.4 in the property (c). Then, $N = Q_0 \supset Q_p$ and $G^p(Q_p) = I_p = N$. By the property (b) there exists in Q_p a fixed point τ_* of the map $F = G^p$. It is evident that inequality $G^k(\tau_*) \neq \tau_*$ holds for $k = 1, 2, \ldots, p-1$ because in the opposite case we would have $G^{k-1}(\tau_*) = b$, which contradicts the fact that $G^{k+1}(\tau_*) = G^2(b) = d \notin N$. $\qquad\square$

4.3 Rotation interval and frequency locking regions

If T is a continuous non-decreasing function satisfying the condition (4.10) then, according to Theorem 2.6, the existence of periodic points and the type of these points are completely determined by the rotation number

$$\rho(\tau, T) = \lim_{p \to \infty} \frac{T^p(\tau)}{2\pi p} \qquad (4.26)$$

of the map T. The limit (4.26) exists and does not depend on τ.

If T is not monotonic then the problem is more complicated. The limit (4.26) depends on the point τ and it can exist or not. For non-monotonic maps the rotation number (4.26) (if it exists) is a property of the trajectory $\{T^p(\tau)\}$ $p = 0, 1, 2, \ldots$ (but not of the map T). For example, we have $\rho(\tau_0, T) = \frac{n}{m}$ for the trajectory of a periodic point τ_0 of the type n/m.

4.3.1 *Definition and properties*

An important characteristic of non-monotonic map T is the *rotation interval*.

Definition 4.2. The closed interval $\rho[T] = [r_-, r_+]$, where

$$r_- = \min_{\tau} \left(\liminf_{p \to \infty} \frac{T^p(\tau)}{2\pi p} \right), \qquad r_+ = \max_{\tau} \left(\limsup_{p \to \infty} \frac{T^p(\tau)}{2\pi p} \right) \qquad (4.27)$$

is called the rotation interval of the map T satisfying the condition (4.10).

For certain maps (e.g. for an increasing one) the rotation interval reduces to a point $\rho[T] = r_- = r_+$.

The following theorem (see Ref. [7]) is a generalization of Theorem 4.2.

Theorem 4.6. *Let T be a continuous map satisfying the condition (4.10), and let $\rho[T]$ be the rotation interval of this map. If $a \leq b$ and $[a, b] \subset \rho[T]$,* *then there exists a point τ^* such that*

$$\liminf_{p\to\infty} \frac{T^p(\tau^*)}{2\pi p} = a, \qquad \limsup_{p\to\infty} \frac{T^p(\tau^*)}{2\pi p} = b. \qquad (4.28)$$

The proof is omitted.

The rotation interval is determined by rotation numbers of two monotonic maps. For a given function T we define two non-decreasing functions:

$$T_+(\tau) = \max_{x\leq\tau} T(x), \qquad T_-(\tau) = \min_{x\geq\tau} T(x). \qquad (4.29)$$

The inequalities

$$T_-(\tau) \leq T(\tau) \leq T_+(\tau) \qquad (4.30)$$

are satisfied for every τ. The function T_+ is the least monotonic function which bounds T from above, and T_- is the greatest monotonic function which bounds T from below (see Fig.4.5).

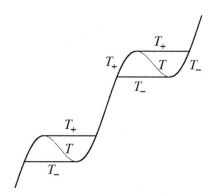

Fig. 4.5   Graphs of the functions (4.29).

For the composition of two functions T and S, the identities

$$(S \circ T)_+ = S_+ \circ T_+ \qquad \text{and} \qquad (S \circ T)_- = S_- \circ T_- \qquad (4.31)$$

are true. Indeed,

$$(S \circ T)_+(\tau) = \max_x \left\{ S\big(T(x)\big) : x \leq \tau \right\}, \qquad (4.32)$$

$$(S_+ \circ T_+)(\tau) = \max_y \left\{ S(y) : y \leq T_+(\tau) \right\}, \qquad (4.33)$$

and from equality $\{T(x) : x \leq \tau\} = \{y : y \leq T_+(\tau)\}$ (see Fig.4.5) it follows that two expressions (4.32) and (4.33) are identical. The proof of the second identity (4.31) is similar.

For $S = T$, from (4.31) it follows that $(T^2)_+ = (T_+)^2$ and by induction

$$(T^n)_+ = (T_+)^n, \qquad (T^n)_- = (T_-)^n. \qquad (4.34)$$

We will write T_+^n and T_-^n because, by (4.34), the parentheses prove to be superfluous.

For non-decreasing functions T_+ and T_-, the rotation numbers

$$\rho(T_+) = \lim_{p \to \infty} \frac{T_+^p(\tau)}{2\pi p}, \qquad \rho(T_-) = \lim_{p \to \infty} \frac{T_-^p(\tau)}{2\pi p} \qquad (4.35)$$

exist and do not depend on a point τ.

The following theorem determines the basic characteristics of the rotation interval.

Theorem 4.7. *Let T be a continuous map which satisfies (4.10). Then*

(a) *the rotation interval is*

$$\rho[T] = [\rho(T_-), \rho(T_+)], \qquad (4.36)$$

(b) *for coprime numbers n, m, the periodic point of the type n/m exists if and only if $\frac{n}{m} \in \rho[T]$,*

(c) *if a non-monotonic (with respect to τ) map $T(\tau; \mu)$ is a continuous and increasing function of a parameter μ in an interval $[\mu_1, \mu_2]$, then the end-points of the rotation interval $\rho(T_-; \mu)$ and $\rho(T_+; \mu)$ are the Cantor step-functions or constant functions of μ for $\mu \in [\mu_1, \mu_2]$.*

Proof. (a) From (4.30) and (4.34) it follows that

$$T_-^n(\tau) \le T^n(\tau) \le T_+^n(\tau)$$

for each $n = 0, 1, 2, \ldots$ Hence, $\rho(T_-) \le r_- \le r_+ \le \rho(T_+)$. On the other hand, we have the inequality

$$\lim_{n_k \to \infty} \frac{T^{n_k}(x)}{2\pi n_k} \le r_+ = \max_\tau \left[\limsup_{p \to \infty} \frac{T^p(\tau)}{2\pi p} \right] \qquad (4.37)$$

for each subsequence $\{n_k\}$ and for each point x for which the limit exists. For each natural number n let us choose such x_n that $T^n(x_n) = (T^n)_+(x_n)$ and $x_n \in [0, 2\pi]$. In the set $\{x_n : n = 1, 2, 3, \ldots\}$ there exists a subsequence $\{x_{n_k}\}$ which converges to a point x_∞. For $x = x_\infty$ there exists the limit in the left-hand side of (4.37) and it is equal to $\rho(T_+)$. It follows from the equality

$$T^{n_k}(x_\infty) = T_+^{n_k}(x_\infty) + \left(T_+^{n_k}(x_{n_k}) - T_+^{n_k}(x_\infty) \right) + \left(T^{n_k}(x_\infty) - T^{n_k}(x_{n_k}) \right)$$

and from the two estimates: $|T_+^{n_k}(x_{n_k}) - T_+^{n_k}(x_\infty)| \le 2\pi$ (because T_+ is an increasing function) and $|T^{n_k}(x_\infty) - T^{n_k}(x_{n_k})| \le n_k \sup_x |T'(x)| \, |x_\infty - x_{n_k}|$.

So, we get $\rho(T_+) \leq r_+$. Similarly, we can prove that $\rho(T_-) \geq r_-$ and, eventually, (4.36) holds.

(b) If τ_0 is a periodic point of the type n/m of the map T, then for $\tau = \tau_0$ there exists the limit (4.26), and $\rho(\tau_0, T) = \frac{n}{m} \in \rho[T]$.

On the other hand, if $\frac{n}{m} \in \rho[T]$ then by Theorem 4.6 for $a = b = \frac{n}{m}$, there exists a point τ^* such that $\rho(\tau^*, T) = \frac{n}{m}$. The existence of a periodic point of the type n/m follows from Theorem 2.6(b) (Section 2.4.3).

(c) The map $T_+(\tau; \mu)$ is a continuous and non-decreasing function of variable τ and an increasing function of μ. If $T(\tau, \mu)$ is non-monotonic, then there exist two points $\tau_1 \neq \tau_2$ such that $T_+(\tau_1) = T_+(\tau_2)$ and, consequently, $T_+^n(\tau_1) = T_+^n(\tau_2)$ for every natural n. The periodic function $P_+^n(\tau) = T_+^n(\tau) - \tau$ is not a constant function, and hence $T_+(\tau; \mu)$ has T-*property*. All assumptions of Theorem 2.11 are satisfied, and $\rho(T_+, \mu)$ is Cantor's step-function. Similarly, $\rho(T_-, \mu)$ is also Cantor's step-function. □

4.3.2   Selected frequency locking regions

Theorem 4.7 will be used for construction of selected rotation intervals $\rho[T] = [r_-, r_+]$ of the map (4.8) as a function of the parameters μ and a.

First, in the plane of parameters μ, a the set of points will be given for which the end-points $r_- = \rho(T_-)$ and $r_+ = \rho(T_+)$ of the rotation interval are integers. The equality $r_+ = n$ holds if and only if the equation

$$T_+(\tau) = \tau + 2\pi n, \qquad \text{where } T_+(\tau) = \max_{x \leq \tau} \left(x + 2\pi\mu + a \sin x \right) \quad (4.38)$$

has a solution or, equivalently, if $2\pi(n - \mu)$ belongs to the set of values of the continuous periodic function

$$P(\tau; a) = \max_{x \leq \tau} \left(x + a \sin x \right) - \tau. \quad (4.39)$$

If $|a| \leq 1$ then $P(\tau; a) = a \sin \tau$. If $a > 1$ and $\tau \in [\frac{1}{2}\pi, \frac{5}{2}\pi]$ then

$$P(\tau; a) = \begin{cases} a \sin \tau & \text{for } \tau \notin [\tau_1, \tau_2], \\ (\tau_1 - \tau) + a \sin \tau_1 & \text{for } \tau \in [\tau_1, \tau_2], \end{cases} \quad (4.40)$$

where the segment $[\tau_1, \tau_2]$ is determined by

$$T'(\tau_1) = 0, \qquad \frac{\pi}{2} < \tau_1 < \pi < \tau_2 < \frac{5\pi}{2}, \qquad T(\tau_2) = T(\tau_1),$$

(see Fig.4.6) or, equivalently, by

$$a \cos \tau_1 = -1, \qquad \tau_2 + a \sin \tau_2 = \tau_1 + a \sin \tau_1. \quad (4.41)$$

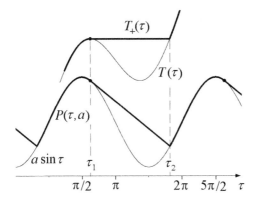

Fig. 4.6   Graphs of functions (4.38) and (4.40).

Hence $\tau_1 = \arccos(-1/a)$, but $\tau_2 = \tau_2(a)$ can be obtained by numerical methods only.

From (4.40) it follows that

$$\max_\tau P(\tau; a) = a,$$

$$\min_\tau P(\tau; a) = \begin{cases} -a & \text{for } \tau_2 \le \tfrac{3}{2}\pi \ (\text{for } a \le a_0), \\ a \sin \tau_2(a) & \text{for } \tau_2 > \tfrac{3}{2}\pi \ (\text{for } a > a_0), \end{cases}$$

where $a_0 = 1.38005...$ is the solution of the equation $\tau_2(a_0) = 3\pi/2$ which can be reduced to

$$a_0 + \sqrt{a_0^2 - 1} + \arccos(-1/a_0) = 3\pi/2.$$

It is not hard to see that $P(\tau; -a) = P(\tau + \pi; a)$ and to obtain a similar result for $a < -1$. Eventually, the equality $r_+ = n$ holds if and only if

$$|a| \ge 2\pi(n - \mu) \ge \begin{cases} -|a| & \text{for } |a| \le a_0, \\ |a| \sin \tau_2(|a|) & \text{for } |a| \ge a_0. \end{cases} \tag{4.42}$$

In a similar way one can determine the set of parameters (μ, a) for which $r_- = n$. This set is given also by (4.42) provided that $2\pi(\mu - n)$ is put in the place of $2\pi(n - \mu)$.

Both sets

$$\{(\mu, |a|) : r_+ = n\} \quad \text{and} \quad \{(\mu, |a|) : r_- = n\} \tag{4.43}$$

are shown in Fig.4.7 for three successive numbers n.

The intersection of two sets (4.43) is called *frequency locking region* with rotation number equal to n. It looks like the shaded domain in Fig.4.7.

Let us notice that it is not necessary to solve numerically the equations (4.41). In the plane of parameters μ, a, the line described by the equation

$$2\pi(n - \mu) = a \sin \tau_2(a) \qquad \text{for } a \geq a_0 = 1.38005... \qquad (4.44)$$

can be expressed by a parametric representation. Let us put $\tau_1 = \gamma - \beta$ and $\tau_2 = \gamma + \beta$ in (4.41) and (4.44). We get

$$a(\cos\gamma \cos\beta + \sin\gamma \sin\beta) = -1,$$
$$a \sin\beta \cos\gamma = -\beta,$$
$$2\pi(n - \mu) = a(\sin\gamma \cos\beta + \cos\gamma \sin\beta).$$

From the first two equations it follows that $\tan\gamma = \beta^{-1} - \cot\beta$ and, eventually,

$$a = \frac{1}{\sin\beta}\sqrt{\beta^2 + (1 - \beta\cot\beta)^2}, \qquad (4.45)$$

$$\mu = n + \frac{1}{2\pi}(\beta + \cot\beta - \beta\cot^2\beta). \qquad (4.46)$$

The equations (4.45) (4.46) are a parametric representation of the line (4.44) with the parameter $\beta \in (1.16556..., \pi)$.

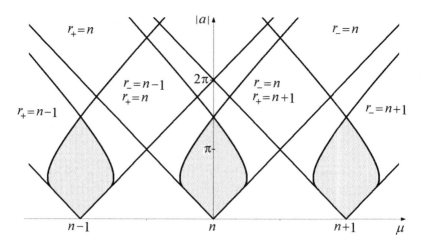

Fig. 4.7   Frequency locking regions with rotation numbers $n-1$, n, $n+1$.

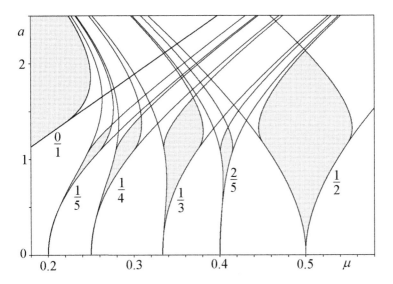

Fig. 4.8   Frequency locking regions of the map (4.8) for several rational values of the rotation number.

For $m > 1$ the regions of points $(\mu, |a|)$ for which $r_- = \frac{n}{m}$ or $r_+ = \frac{n}{m}$, can be obtained by numerical methods only. They are shown in Fig.4.8 for several numbers $\frac{n}{m} = \frac{1}{2}, \frac{1}{3}, \frac{1}{4}, \frac{1}{5}, \frac{2}{5}$.

In the frequency locking regions shaded in the picture, the rotation interval reduces to the rational point $\rho[T] = r_- = r_+ = \frac{n}{m}$. These regions exceed the range $|a| < 1$ where the map T is monotonic (for monotonic maps the rotation interval always reduces to a point).

A graph of the rotation interval $\rho[T]$ of the map (4.8) versus μ is shown in Fig.4.9 for the fixed $a = 1.05$. Both end-points of rotation intervals are the Cantor step-functions of μ. For several intervals of μ the rotation interval reduces to the rotation numbers: $\frac{1}{5}, \frac{3}{14}, \frac{2}{9}, \frac{3}{13}, \frac{4}{17}, \frac{1}{4}$. It means that in the plane of parameters μ, a the line $a = 1.05$ intersects respective frequency locking regions.

If $\rho[T] = \frac{n}{m}$ then for all trajectories there exist the limits (4.26) and, for the given phase-locked loop, the quotient of the input signal frequency and the average output signal frequency (or the quotient of the "average period" of the output signal and the period of the input signal) is the constant value equal to $\frac{n}{m}$ independently of an initial condition. The system "tracks" the frequency of the input signal. It does not mean, however, that a stable

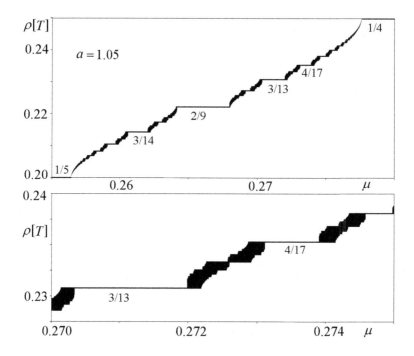

Fig. 4.9   Rotation interval of the map (4.8) versus μ.

periodic output signal of the type n/m exists.

4.3.3   *Application to the map (4.7)*

The frequency locking regions of the map (4.7) are shown in Fig.4.10 (where $\mu = \omega/\Omega$) for several values of the rotation number.

Similarly as in Section 4.3.2, the boundaries of these regions are given by analytic formulas only for $\rho = 1, 2, 3...$ For $\rho = \frac{1}{2}, \frac{3}{2}, \frac{5}{2}$, the frequency locking regions were obtained by numerical methods. Let $[r_-, r_+]$ denote a rotation interval of the map (4.7). The domains of the parameters μ, A, where r_- or r_+ take the values 1, 2, 3, are marked in the figure.

For each natural number n in the domain $|\mu - n| < nA$ there exists a periodic point of the type $n/1$, stable or unstable.

In the domains bounded from below by straight lines $|\mu - n| = nA$, shown for $n = 1, 2, 3$, and from above by dotted lines, there exist stable periodic points of the types $1/1$, $2/1$ and $3/1$ respectively.

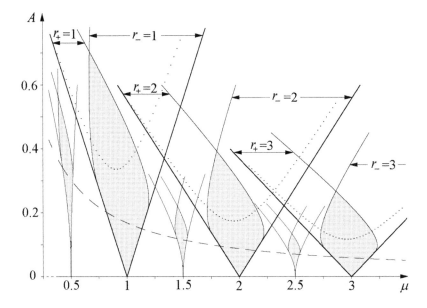

Fig. 4.10   Frequency locking regions of the map (4.7).

Below the dashed line the map (4.7) is an increasing function of τ, and all Arnold's tongues are disjoint in this domain.

4.4   Stable orbits, hold-in regions

If a point (μ, A) belongs to a frequency locking region then we have fixed relation between frequencies of the input and output signals but, in general, there is no relation between phases of the both signals. The phase difference between input and output signals does not change in steady-state in such domains of parameters (hold-in regions), where the map T has a stable periodic orbit.

4.4.1   *Stability of periodic points*

Let τ_0 be a periodic point of the type n/m of a smooth map T which satisfies the condition (4.10), and let $\{\tau_0, \tau_1, \ldots \tau_{m-1}\}$ $(mod\, 2\pi)$ be a periodic orbit of the point τ_0. Every point of the periodic orbit is a fixed point of the map $G(\tau) = T^m(\tau) - 2\pi n$.

The number

$$d \overset{\text{def}}{=} G'(\tau_0) \equiv T'(\tau_0) \cdot T'(\tau_1) \cdot T'(\tau_2) \cdot \ldots \cdot T'(\tau_{m-1}) \qquad (4.47)$$

is called the *multiplier of the periodic orbit* (see Ref.[15]). It depends on all points of the orbit.

If $T'(c) = 0$ then c is called a critical point, and $T(c)$ is called a critical value of the map T.

Definition 4.3. A periodic point τ_0 and corresponding periodic orbit is called

 – stable (attracting) if $|d| < 1$,
 – superstable (superattracting) if $d = 0$,
 – unstable (repelling) if $|d| > 1$,
 – neutral (indifferent) if $|d| = 1$.

Let the multiplier d be different from $0, +1, -1$. As long as the terms of the sequence $\{x, G(x), G^2(x), \ldots\}$ belong to a sufficiently small neighborhood of an arbitrary point τ_p of the periodic orbit, the equality $G^k(x) - \tau_p = d^k(x - \tau_p) + O\big((x - \tau)^2\big)$ holds. So, the sequence $\{G^k(x) - \tau_p\}$ is close to a geometric progression with the ratio d. Therefore, the orbit is called attracting for $|d| < 1$ and repelling for $|d| > 1$. For superstable orbits the sequence $\{G^k(x) - \tau_p\}$ tends to zero faster than a geometric progression with an arbitrary small ratio $|d| > 0$. A superstable orbit always contains a critical point. For a neutral orbit the convergence of the sequence $\{G^k(x) - \tau_p\}$ depends on the $\text{sgn}(x - \tau_p)$ and on higher derivatives of the map T at the points of the orbit.

4.4.2   *Stable periodic points of the type $n/1$ and $n/2$*

We will be concerned with regions of parameters μ, a for which the map (4.8) has stable periodic points of the types $n/1$ and $n/2$.

Graph of the function (4.8) for selected values of parameters is shown in Fig.4.11. For these values of parameters there exist a stable fixed point τ^* of the type $1/1$ and a stable periodic orbit $\{\tau_0, \tau_1\}$ of the type $0/2$.

Let us denote $\xi = 2\pi(\mu - n)$. If τ_0 is a periodic point of the type $n/1$ with multiplier $d = T'(\tau_0)$ then the equalities

$$a \sin \tau_0 = -\xi, \qquad 1 + a \cos \tau_0 = d \qquad (4.48)$$

are fulfilled. The periodic point τ_0 exists if and only if $|a| \geq |\xi|$. Moreover, a stable periodic point exists if and only if $d \in (-1, +1)$ i.e.

$$|\xi| < |a| < \sqrt{4 + \xi^2}. \qquad (4.49)$$

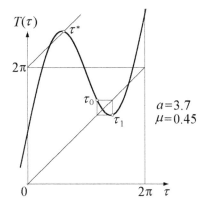

Fig. 4.11   Graph of the function (4.8) for selected values of parameters.

Fig.4.12 shows a few regions of existence of stable periodic points of the types $n/1$ and an enlarged fragment of one region with two families of lines:
– the family of straight lines $\xi = -a\sin\tau_0$, with $\tau_0 = const.$
– the family of hyperbolae $a^2 - \xi^2 = (1-d)^2$, with $d = const.$

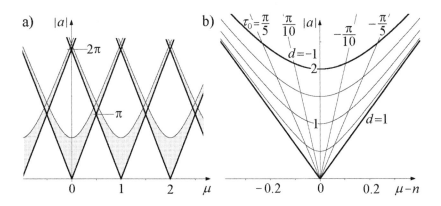

Fig. 4.12   Regions of existence of stable periodic points τ_0 of the types $n/1$, and an enlarged fragment of one region with the lines $\tau_0 = const.$ and $d = const.$

Now we will determine the range of parameters μ, a for which the map (4.8) has a stable periodic orbit $\{\tau_0, \tau_1\}$ of the type $n/2$ with a multiplier $d \in (-1, +1)$. For this orbit we have

$$T(\tau_0) = \tau_1, \qquad T(\tau_1) = \tau_0 + 2\pi n, \qquad T'(\tau_0)T'(\tau_1) = d$$

and $\tau_0 - \tau_1 \neq \pi n$ for even numbers n. The points τ_0, τ_1 can be determined from the equations

$$a \sin \tau_0 = \tau_1 - \tau_0 - 2\pi\mu,$$
$$a \sin \tau_1 = \tau_0 - \tau_1 + 2\pi(n - \mu),$$
$$(1 + a \cos \tau_0)(1 + a \cos \tau_1) = d. \tag{4.50}$$

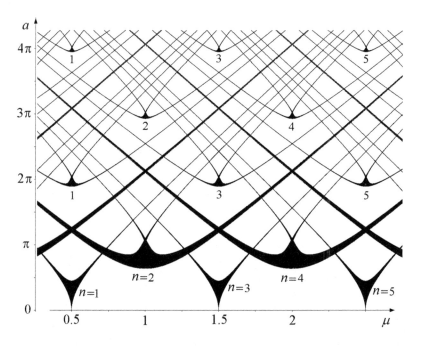

Fig. 4.13   Regions of existence of stable periodic orbits of the types $n/2$. The number n is shown for each region.

It is convenient to introduce the notations

$$\tau_1 - \tau_0 - n\pi = 2\beta, \qquad \tau_1 + \tau_0 = 2\gamma, \qquad 2\pi\mu - n\pi = \xi$$

and write (4.50) in the form

$$a \sin \gamma \cos(\beta + n\pi/2) = -\xi,$$
$$a \cos \gamma \sin(\beta + n\pi/2) = -2\beta, \tag{4.51}$$
$$1 + 2a \cos \gamma \cos(\beta + n\pi/2) + a^2 \cos^2 \gamma - a^2 \sin^2(\beta + n\pi/2) = d.$$

Eliminating γ we obtain the solution of equations (4.51) with respect to ξ^2 and a^2,

for even n:
$$\xi^2 = \cot^2 \beta \Big((1 - 2\beta \cot \beta)^2 - d \Big),$$
$$a^2 = (1 + \cot^2 \beta) \Big((1 - 2\beta \cot \beta)^2 + 4\beta^2 - d \Big),$$
(4.52)

for odd n:
$$\xi^2 = \tan^2 \beta \Big((1 + 2\beta \tan \beta)^2 - d \Big),$$
$$a^2 = (1 + \tan^2 \beta) \Big((1 + 2\beta \tan \beta)^2 + 4\beta^2 - d \Big).$$
(4.53)

For a fixed d, the formulae (4.52) and (4.53) are parametric representations (with a parameter β) of the lines lying in the plane μ, a, where $\mu = \frac{1}{2\pi}\xi + \frac{n}{2}$. For every point of these lines there exists a periodic orbit $\{\tau_0, \tau_1\} \, mod \, 2\pi$ with a multiplier d. The orbits are stable for $d \in (-1, +1)$ and superstable for $d = 0$. Neutral orbits exist for $d = \pm 1$.

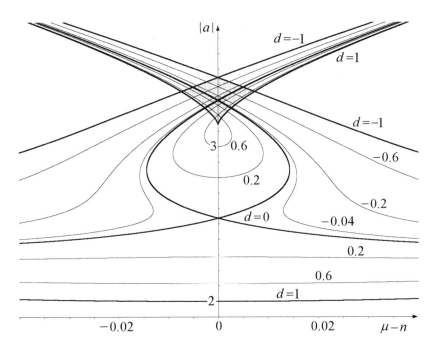

Fig. 4.14   The lines of existence of stable periodic orbits of the type $2n/2$ with multipliers d.

In the plane of the parameters μ, a the regions of existence of stable periodic orbits $\{\tau_0, \tau_1\}$ of the types $n/2$ are shown in Fig.4.13 for several integers n. These regions are symmetric with respect to the lines $\mu = n$ and $\mu = n + \frac{1}{2}$ (or equivalently, with respect to $\xi = 0$ and $\xi = \pi$).

An enlarged fragment of Fig.4.13 and, additionally, the family of lines $d = const.$ are shown in Fig.4.14. Stable periodic orbits of the type $2n/2$ exist in the region bounded by the lines $d = 1$ and $d = -1$. The lower bound $d = 1$ of the region (described by equation $a^2 - \xi^2 = 4$) is equal to the upper bound of the region of existence of stable periodic points of the type $n/1$. Crossing over this line is connected with the period doubling bifurcation of the stable periodic orbit (see Sec.4.6.2). In the shaded domain there exist two stable orbits of the type $2n/2$. On the line $d = 0$ a periodic orbit contains one critical point c_1 or c_2. At the point $\mu - n = 0$, $a = 2.5365...$ the orbit $\{c_1, c_2\}$ contains both critical points.

4.4.3   *Attractive set of a fixed point*

Let τ_0 be a stable periodic point of the type n/m of a smooth map T which satisfies the condition (4.10). Then, $G(\tau_0) = \tau_0$ and $|G'(\tau_0)| < 1$, where $G(\tau) = T^m(\tau) - 2\pi n$.

Definition 4.4. The subset

$$A(\tau_0) = \left\{ x : \lim_{k \to \infty} G^k(x) = \tau_0 \right\} \tag{4.54}$$

of the real line R is called the attractive set of the stable fixed point τ_0 of the map G (or the attractive set of the periodic point τ_0 of the map T).

The attractive set of the stable fixed point τ_0 of the map G is usually a union of open intervals. One of them includes the point τ_0.

Definition 4.5. The maximal open interval $A^0(\tau_0)$ which satisfies the condition $\tau_0 \in A^0(\tau_0) \subset A(\tau_0)$ is called the immediate attractive interval of the fixed point τ_0.

If we know the graph of the function $G(\tau)$ then determining the immediate attractive interval of a stable fixed point is not difficult.

Theorem 4.8. *Let $G(\tau_0) = \tau_0$ and $|G'(\tau_0)| < 1$. An open interval (s_1, s_2) containing the point τ_0 is the immediate attractive interval of the fixed point τ_0 if and only if it is the smallest interval the end-points of which fulfil one* *of the four following conditions (see* Fig.4.15)*:*

a) $G(s_1) = s_1 \quad G(s_2) = s_2,$

b) $G(s_1) = s_1 \quad G(s_2) = s_1,$

c) $G(s_1) = s_2 \quad G(s_2) = s_2,$ \hfill (4.55)

d) $G(s_1) = s_2 \quad G(s_2) = s_1.$

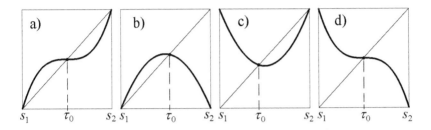

Fig. 4.15   Graphs of the functions $G(x)$ in the immediate attractive intervals of stable fixed points τ_0.

Proof. If the assumptions of Theorem 4.8 are satisfied then for every natural number m the function G^m maps the interval (s_1, s_2) into itself and also satisfies one of the conditions (4.55) (with G^m instead of G). Moreover,

$$G^m(\tau_0) = \tau_0, \quad \text{and} \quad \left| \frac{dG^m}{d\tau}(\tau_0) \right| \equiv |G'(\tau_0)|^m < 1. \quad (4.56)$$

In a sufficiently small neighborhood of τ_0 we have

$$G^m(x) > x \text{ for } x < \tau_0 \quad \text{and} \quad G^m(x) < x \text{ for } x > \tau_0$$

These inequalities can be extended to the whole interval (s_1, s_2):

$$s_2 > G^m(x) > x \quad \text{for} \quad x \in L \equiv (s_1, \tau_0), \quad (4.57)$$

$$s_1 < G^m(x) < x \quad \text{for} \quad x \in R \equiv (\tau_0, s_2). \quad (4.58)$$

Indeed, in the opposite case we would get a contradiction to the assumption that (s_1, s_2) is the smallest interval where the function G^m fulfils one of the conditions (4.55).

Now we will show that for any $x \in (s_1, s_2)$ the sequence

$$\{x, G(x), G^2(x), G^3(x), \ldots\} \quad (4.59)$$

converges to τ_0. For this purpose, we form two subsequences

$$\{x_{l0}, x_{l1}, x_{l2}, \ldots\}, \qquad \{x_{r0}, x_{r1}, x_{r2}, \ldots\}. \quad (4.60)$$

The first of them contains those terms of the sequence (4.59) which belong to the interval L, and the second contains those terms which belong to R.

If $G^k(x) \in L$ and $G^{k+m}(x) \in L$ for two natural numbers k, m, then from (4.57) it follows that $s_1 < G^k(x) < G^{k+m}(x) < \tau_0$. Therefore, the first sequence (4.60) is increasing and bounded from above by τ_0. Similarly, it can be shown that the second sequence is decreasing and bounded from below by τ_0.

There are the following possibilities:

1. Both sequences (4.60) have a finite number of terms. Then, there exists N such that $G^k(x) = \tau_0$ for $k \geq N$.

2. Exactly one sequence (4.60) has an infinite number of terms. Then, its limit x_∞ satisfies the equation $G(x_\infty) = x_\infty$.

3. Both sequences (4.60) have an infinite number of terms. Then, their limits $x_{l\infty}$ and $x_{r\infty}$ satisfy the equations $G(x_{l\infty}) = x_{r\infty}$ and $G(x_{r\infty}) = x_{l\infty}$.

From (4.57)–(4.58) it follows that in the interval (s_1, s_2) exactly one point τ_0 satisfies the equations $G(x) = x$ and $G^2(x) = x$. So, the sequence (4.59) converges to τ_0. It means that $(s_1, s_2) \subset A^0(\tau_0)$. The points s_1 and s_2 do not belong to $A^0(\tau_0)$, and eventually we get $(s_1, s_2) = A^0(\tau_0)$.   □

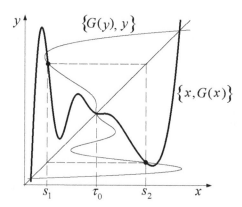

Fig. 4.16   A graphic method of determining the immediate attractive interval.

If an immediate attractive interval satisfies the condition (4.55d), then the points s_1, s_2 can be obtained by the following graphic method (see Fig.4.16). Let us consider two graphs $\{x, G(x)\}$ and $\{G(y), y\}$ of the maps

G and G^{-1} in the xy-plane. One of them is a mirror reflection of the other with respect to the straight line $x = y$. If $(s_1, s_2) \times (s_1, s_2)$ is the largest open square containing only one intersection point (τ_0, τ_0) of both graphs then (s_1, s_2) is the immediate attractive interval of the point τ_0.

If $A^0(\tau_0)$ is the immediate attractive interval of the stable fixed point τ_0 then

$$A(\tau_0) = \bigcup_{k=0}^{\infty} \left\{ x : G^k(x) \in A^0(\tau_0) \right\} \tag{4.61}$$

is the attractive set of the point τ_0.

Theorem 4.9. *An attractive set $A(\tau_0)$ is either the immediate attractive interval $A^0(\tau_0)$ or a union of an infinite number of open disjoint intervals.*

Proof. Let A^1 be the preimage of the interval $A^0 = A^0(\tau_0)$ without A^0:

$$A^1 = \left\{ x : x \notin A^0, \, G(x) \in A^0 \right\}. \tag{4.62}$$

There are two possibilities:

1) If the set A^1 is empty, then $A(\tau_0) = A^0$.

2) If A^1 is not empty, then it is an open interval or a union of open intervals. Since G is a continuous function which maps R onto itself, the preimage $A^2 = \{x : G(x) \in A^1\}$ of the set A^1 is not empty. It is an open interval or a union of open intervals. Moreover, A^2 is disjoint from A^1 and A^0. In fact, let us suppose that there exists $x \in A^2 \cap A^1$. Then $G(x) \in A^1 \cap A^0 = \emptyset$ and, consequently, $A^2 \cap A^1 = \emptyset$. Similarly, if there exists $x \in A^2 \cap A^0$, then $G(x) \in A^1 \cap A^0 = \emptyset$ and, consequently, $A^2 \cap A^0 = \emptyset$. By induction, we can prove that for every $p = 2, 3, 4, \ldots$ the set

$$A^p = \left\{ x : G(x) \in A^{p-1} \right\} \equiv \left\{ x : G^{p-1}(x) \in A^1 \right\} \tag{4.63}$$

is a non-empty open interval or a union of open intervals disjoint from the sets $A^0, A^1, A^2, \ldots, A^{p-1}$. Eventually, the attractive set $A(\tau_0) = \bigcup_{p=0}^{\infty} A^p$ is a union of an infinite number of open intervals. $\square$

Corollary 4.3. *If G is an increasing function and $G(x + 2\pi) = G(x) + 2\pi$, then the attractive set $A(\tau_0)$ of a stable fixed point is an open interval (see Fig.4.15a).*

If G is not an increasing function, and if the immediate attractive interval $A^0(\tau_0)$ of a fixed point τ_0 satisfies the condition (4.55 b,c or d), then the attractive set $A(\tau_0)$ is the union of an infinite number of open intervals.

4.4.4   *Attractive set of a stable periodic orbit*

If τ_0 is a periodic point of the type n/m of the map T, then for every natural number p, the points $\tau_0 + 2\pi p$ and $T^p(\tau_0)$ are also periodic points of the same type. It is not hard to see that the equalities

$$A(\tau_0 + 2\pi p) = A(\tau_0) + 2\pi p \equiv \{x + 2\pi p : x \in A(\tau_0)\},$$

$$A\big(T^p(\tau_0)\big) = T^p\big(A(\tau_0)\big) \equiv \{T^p(x) : x \in A(\tau_0)\}$$

are satisfied for every $p = 1, 2, 3, \ldots$.

Definition 4.6. Let $\{\tau_0, \tau_1, \tau_2, \ldots, \tau_{m-1}\} \, mod \, 2\pi$ be a stable periodic orbit of the map T. The subset

$$A_{\text{orb}}(\tau_0) = \bigcup_{i=0}^{m-1} A(\tau_i) \, mod \, 2\pi \tag{4.64}$$

of the circle S is called the attractive set of this orbit.

Example 4.2. The graph of the function

$$T(\tau) = \tau + 2.4 \, \frac{2\pi}{1 + 0.21 \sin \tau} \tag{4.65}$$

and graph of $T(\tau) \, mod \, 2\pi$ are shown in Fig.4.17.

The map T has two $(mod \, 2\pi)$ stable periodic points: $\tau'' = 1.2609517$ of the type $2/1$, and $\tau''' = 5.0222336$ of the type $3/1$. Using Theorem 4.8 and the construction described in the proof of Theorem 4.9, the attractive sets $A_{\text{orb}}(\tau'')$ and $A_{\text{orb}}(\tau''')$ of the stable orbits (fixed points) are shown. Both these attractive sets are unions of an infinite number of open intervals. Between any two intervals belonging to $A_{\text{orb}}(\tau'')$ there exists an interval which belongs to $A_{\text{orb}}(\tau''')$, and vice versa.

It can be shown that mes $A_{\text{orb}}(\tau'') +$ mes $A_{\text{orb}}(\tau''') = 2\pi$. In the complement of the set $A_{\text{orb}}(\tau'') \cup A_{\text{orb}}(\tau''')$ with respect to the interval $[0, 2\pi]$ there exists an infinite set of unstable periodic orbits and a set of non-periodic trajectories which satisfies Theorem 4.4 and which has the cardinality of continuum.

Let the discrete-time phase-locked loop DPLL described in Section 4.1 by (4.5)–(4.6) have the input signal $U_{\text{inp}}(\omega t) = AV \sin \omega t$, with $A = 0.21$, $\frac{\omega}{\Omega} = 2.4$, and the output signal $U_{\text{out}}\big(\theta(t)\big) = \sin \theta(t)$. The phase $\theta(t)$ depends on the initial value $t_0 = \tau_0/\omega$ (where $\theta(t_0) = 0$). The input signal and two stable periodic output signals are shown in Fig.4.18. If $\omega t_0 \in A_{\text{orb}}(\tau'')$ then $U_{\text{out}}\big(\theta(t)\big)$ tends to the first stable periodic signal, and if $\omega t_0 \in A_{\text{orb}}(\tau''')$ then it tends to the second one for $t \to \infty$.

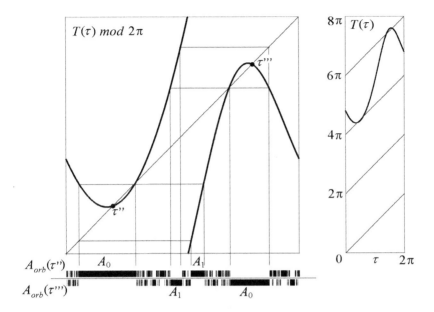

Fig. 4.17   Construction of attractive sets of two stable fixed points τ'' and τ''' .

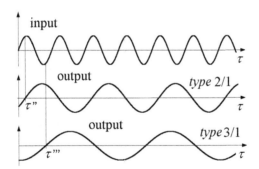

Fig. 4.18   Example of an input signal and two stable output signals of the types $2/1$ and $3/1$ for DPLL system described by (4.65).

4.5   The number of stable orbits

Let the DPLL system be described by the map (4.8). In this section we are dealing with such a region of parameters μ, a where the output signal frequency is n-times less then the input signal frequency, i.e. the frequency locking region with rotation number equal to n. In this region we can find subregions in which the number of stable periodic orbits is:

1) at most two,
2) at most one,
3) exactly one.

Evidently, all these stable orbits (if they exist) are of the type nk/k for a natural number k.

In general, the number of stable periodic orbits of a continuous map T satisfying the condition $T(\tau + 2\pi) = T(\tau) + 2\pi$ can take values from zero to infinity.

Example 4.3. The map $T(\tau) = \tau + P(\tau)$, where $P(\tau + 2\pi) = P(\tau)$, $P(0) = 0$ and

$$P(\tau) = 0.0025\,\tau^3(2\pi - \tau)\sin(4\pi^2/\tau) \quad \text{for} \quad \tau \in (0, 2\pi)$$

has everywhere a positive continuous derivative $T'(\tau) \in (0, 2)$. In the interval $[0, 2\pi)$ the map T has an infinite number of stable fixed points $\tau_k = \frac{2\pi}{k}$ for $k = 2, 3, 4, \ldots$ with the attractive sets $A_{\text{orb}}(\tau_k) = (\frac{4\pi}{2k+1}, \frac{4\pi}{2k-1})$. Moreover, it has an infinite number of unstable fixed points and one neutral fixed point $\tau = 0$.

Suppose that we have found m stable periodic orbits of the map T and we know their attractive sets. If the union of these attractive sets has the measure equal to 2π then there are no more stable orbits, but if it is less than 2π it does not mean that there exists one more stable orbit.

4.5.1 *Schwarzian derivative*

For the map with negative Schwarzian derivative (see Refs. [15], [40]) the number of stable orbits can be estimated from above. Let T satisfy the condition (4.10) and let its third derivative be continuous for all τ.

Definition 4.7. The Schwarzian derivative $\mathcal{S}_d T(\tau)$ of a function T at the point τ is defined by:

$$\mathcal{S}_d T(\tau) = \frac{T'''(\tau)}{T'(\tau)} - \frac{3}{2}\left(\frac{T''(\tau)}{T'(\tau)}\right)^2. \tag{4.66}$$

We say that a function T has negative Schwarzian derivative if $\mathcal{S}_d T(\tau) < 0$ for all τ such that $T'(\tau) \neq 0$.

Example 4.4. The function $T(\tau) = \tau + 2\pi\mu + a\sin\tau$ has negative Schwarzian derivative if and only if $|a| > 1$.

Lemma 4.1. *If a function T has negative Schwarzian derivative, then*
(a) T^n *has negative Schwarzian derivative also for $n = 2, 3, 4, \ldots$*
(b) $|T'|$ *has no positive local minima.*

Proof. (a) It is not hard to see that the Schwarzian derivative of the composite function $F\big(T(\tau)\big)$ is expressed by

$$\mathcal{S}_d(F \circ T)(\tau) = \big(T'(\tau)\big)^2 \mathcal{S}_d F(x) + \mathcal{S}_d T(\tau), \quad \text{where} \quad x = T(\tau). \quad (4.67)$$

So, if both functions F and T have negative Schwarzian derivatives, then their composition has also negative Schwarzian derivative. Replacing F by the functions $T, T^2, T^3, \ldots$ we get the property (a).

(b) It is easy to check that

$$\mathcal{S}_d T(\tau) = -2\sqrt{|T'(\tau)|} \, \frac{d^2}{d\tau^2} \left(\frac{1}{\sqrt{|T'(\tau)|}} \right). \quad (4.68)$$

Suppose that the function $|T'(\tau)|$ has its local minimum at a point τ_0 and $T'(\tau_0) \neq 0$. Then the function $H(\tau) = 1/\sqrt{|T'(\tau)|}$ has its derivative and local maximum at the point τ_0. Hence, $H'(\tau_0) = 0$ and $H''(\tau_0) \leq 0$. By (4.68) we have $\mathcal{S}_d T(\tau_0) \geq 0$. So, we get a contradiction with the assumption that T has negative Schwarzian derivative. The lemma is proved. $\square$

Theorem 4.10 (Singer). *Let the map T satisfying the condition (4.10) has negative Schwarzian derivative. In the attractive set of every stable periodic orbit of the map T there is at least one critical point of T.*

Proof. We will show that the map $G(\tau) = T^m(\tau) - 2\pi n$ with negative Schwarzian derivative has a critical point τ_* in the immediate attractive interval (s_1, s_2) of a stable fixed point $\tau_0 = G(\tau_0)$. This interval is completely characterized by Theorem 4.8. Let us consider each of the four cases described by (4.55).

(a) At the points s_1, τ_0, s_2 there is $G'(s_1) \geq 1$, $G'(\tau_0) < 1$, $G'(s_2) \geq 1$. So the function G' attains a minimum at a point $x \in (s_1, s_2)$. If $G'(x) \leq 0$ then there exists a point $\tau_* \in (s_1, s_2)$ such that $G'(\tau_*) = 0$. So τ_* is a critical point of the map G. But if $0 < G'(x) < 1$, then the function $|G'|$ attains at the point x the local positive minimum which, by the lemma 4.1, contradicts the assumption that G has negative Schwarzian derivative.

(b) and (c) The existence of a critical point (a local maximum or a local minimum respectively) of the map G in the interval (s_1, s_2) is obvious.

(d) The interval (s_1, s_2) is the immediate attractive interval also for the stable fixed point τ_0 of the map G^2. This map has negative Schwarzian derivative and satisfies the conditions $G^2(s_1) = s_1$ and $G^2(s_2) = s_2$. Moreover, $(G^2)'(s_1) \geq 1$, $(G^2)'(s_2) \geq 1$ and $(G^2)'(\tau_0) < 1$. In part (a) it was shown that there exists a point $\tau_* \in (s_1, s_2)$ that $G'(\tau_*)G'(G(\tau_*)) = 0$. At least one of two points either τ_* or $G(\tau_*)$ is a critical point of the map G.

Let $\{\tau_0, \tau_1, \ldots, \tau_{m-1}\}$ $(mod\ 2\pi)$ be a stable periodic orbit of the type n/m of the map T. The attractive set $A_{\mathrm{orb}}(\tau_0)$ contains the immediate attractive intervals $A^0(\tau_k)$, $k = 0, 1, 2, \ldots, m - 1$ of all fixed points τ_k of the map $G(\tau) = T^m(\tau) - 2\pi n$ which belong to the orbit. If T has negative Schwarzian derivative then so does G. We have shown above that in $A^0(\tau_0) = (s_1, s_2)$ there exists a point τ_* such that

$$G'(\tau_*) \equiv T'(\tau_*) \cdot T'(T(\tau_*)) \cdot T'(T^2(\tau_*)) \cdot \ldots \cdot T'(T^{m-1}(\tau_*)) = 0.$$

So, there exists a number $0 \leq k \leq m - 1$ such that $c = T^k(\tau_*)$ satisfies the condition $T'(c) = 0$. The point c is the critical point of the map T and $c \in A^0(\tau_k) \subset A_{\mathrm{orb}}(\tau_0)$.   □

Corollary 4.4. *The number of stable orbits of the map T which satisfies the condition* (4.10) *and has negative Schwarzian derivative is not greater than the number of critical points of the map T in the interval* $[0, 2\pi)$. *For example, the map* (4.8) *for* $|a| > 1$ *has no more than two stable orbits.*

Sometimes we are interested in the number of stable orbits occurring in a fixed interval.

Corollary 4.5. *Let T maps an interval $[a, b]$ into itself and has negative Schwarzian derivative in this interval. A stable periodic orbit of the map T lying in the interval $[a, b]$ attracts at least one critical point of the map T or at least one of the end-points a, b.*

Proof.   Let $\{\tau_0, \tau_1, \ldots, \tau_{m-1}\}$ be a stable periodic orbit (of the type $0/m$) of the map T in an interval $[a, b]$. Let $A^0(\tau_i)$, for $i = 0, 1, \ldots, m-1$ denote the immediate attractive interval of a stable fixed point τ_i of the map $G(\tau) = T^m(\tau)$. If the points a and b are not attracted to this orbit, then the union $A^0_{\mathrm{orb}} = \bigcup_{i=0}^{m-1} A^0(\tau_i)$ belongs to the interval (a, b). From the last part of the proof of Theorem 4.10 it follows that the set A^0_{orb} contains a critical point of the map T.   □

4.5.2   Application to the map $T(\tau) = \tau + 2\pi\mu + a\sin\tau$

We will use the Singer theorem (especially Corollary 4.5) to determine the number of stable orbits of the map (4.8) in such a region of parameters μ, a

where the rotation interval reduces to the point $\rho[T] = n$. The boundary of this region were determined in Section 4.3.2 and shown in Fig.4.7. In this region the DPLL system divides by n the frequency of the input signal.

Let z denote such an unstable fixed point of the map

$$G(\tau) = T(\tau) - 2\pi n = \tau + 2\pi(\mu - n) + a\sin\tau \qquad (4.69)$$

for which $G'(z) > 1$. If $\rho[T] = n$, then (4.69) maps the interval $(z, z + 2\pi)$ onto itself (see Fig.4.19).

If $\min\limits_{\tau} G'(\tau) > -1$, i.e. if $|a| < 2$, then in $(z, z + 2\pi)$ there exists exactly one fixed point of the map (4.69). This point is stable and $(z, z + 2\pi)$ is its immediate attractive interval.

For $|a| > 2$, there exist two critical points $c_1 < c_2$ in the interval $(z, z + 2\pi)$ and $G''(c_1) < 0$, $G''(c_2) > 0$. It is easily seen that for every $\tau \in (z, z + 2\pi)$, there exists such a number k that $G^k(\tau) \in I \equiv [G(c_2), G(c_1)]$. So, the closed interval I attracts all points of the interval $(z, z + 2\pi)$. Stable periodic orbits can exist in I only. It is necessary to distinguish three cases

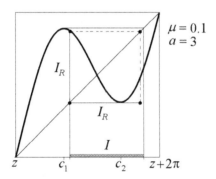

$$\mu = 0.1$$
$$a = 3$$

Fig. 4.19   Construction of the attractive invariant set $I_R = G(I_R)$ of the map (4.69).

1) Both critical points c_1 and c_2 belong to the interval I. Then, there exist at most two stable orbits in I.

2a) The interval I contains only one critical point c_1. Then the interval $I_L = [G^2(c_1), G(c_1)]$ attracts all points of the interval $(z, z + 2\pi)$ and $c_1 \in I_L = G(I_L)$. The critical point c_1 and the end-points of the interval I_L lie on the same trajectory. So, by Corollary 4.5, there exists at most one stable periodic orbit in I_L (and also in $(z, z + 2\pi)$).

2b) The interval I contains only one critical point c_2. Then the interval $I_R = [G(c_2), G^2(c_2)]$ attracts all points of the interval $(z, z + 2\pi)$ and there exist at most one stable periodic orbit in I_R (see Fig.4.19).

3) The interval I does not contain critical points, and G is a decreasing function in I. Then, there exists in I either a stable fixed point τ_* attracting all points of the interval $(z, z + 2\pi)$ or the stable orbit $\{\tau_0, \tau_1\}$ attracting all points of the interval $(z, z + 2\pi)$ except one unstable fixed point τ_*. The line $a^2 - \xi^2 = 4$, where $\xi = 2\pi(\mu - n)$, separates two above cases in the plane of parameters (line L_3 in Fig.4.20).

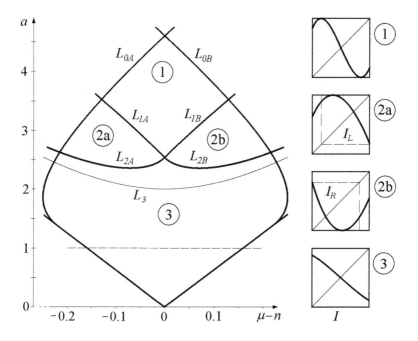

Fig. 4.20 Frequency locking region $\rho[T] = n$ bounded by the lines L_{0A}, L_{0B} and by the line $a = 2\pi|\mu - n|$. The number of stable periodic orbits is: at most two in domain (1), at most one in domains (2a) and (2b), exactly one (of the type $n/1$ below the L_3 and of the type $2n/2$ above the L_3) in domain (3).

Fig.4.20 shows in the μa-plane the lines L_{1A}, L_{1B} separating the domains in which the cases 1 and 2a, 2b occur. They are the loci of points (μ, a) for which both critical points are adjacent points of the same trajectory. Hence,

$$G(c_1) = c_2 \text{ and } z < G(c_2) < c_1, \quad \text{or} \quad G(c_2) = c_1 \text{ and } c_2 < G(c_1) < z + 2\pi.$$

Using (4.69) it is easy to check that the parametric representation of the lines L_{1A}, L_{1B} is

$$\mu = n \pm \frac{1}{2\pi}(2\alpha - \tan\alpha), \qquad |a| = \frac{1}{\cos\alpha}, \qquad \alpha \in [1.166..., 1.28...]. \quad (4.70)$$

On the lines L_{2A}, L_{2B} separating the domains where the cases 2a, 2b and 3 occur the conditions

$$G^2(c_1) = c_1 \text{ and } G(c_1) < c_2, \qquad \text{or} \qquad G^2(c_2) = c_2 \text{ and } G(c_2) > c_1$$

are satisfied. It is easy to check that the lines L_{2A}, L_{2B} overlap the lower branches of the curves where superstable orbits $\{\tau_0, \tau_1\}$ of the period 2 occur. According to (4.52) the parametric representation of the lines L_{2A} and L_{2B} takes the form

$$\mu = n \pm \tfrac{1}{2\pi} \cot\beta(1 - 2\beta\cot\beta),$$

$$|a| = (\sin\beta)^{-1} \sqrt{4\beta^2 + (1 - 2\beta\cot\beta)^2}, \qquad \beta \in [0.55..., 1.166...]. \quad (4.71)$$

It is necessary to emphasize that Singer's Theorem gives information about the maximal number of stable orbits only. In particular cases the number of these orbits may be smaller. For example, if $\mu - n = 1/4\pi$ and $a = 3.010...$ (case 2b) then there is no stable orbit because the trajectory of the critical point c_2 hits the unstable fixed point $\tau_* = G^3(c_2)$. Similarly, for $\mu = n$ and $a = 3.901...$ (case 1), stable orbits do not exist because the trajectories of both critical points hit the unstable fixed point $\tau_* = G^2(c_1) = G^2(c_2) = 0$.

4.6 Bifurcations of periodic orbits

Let the map T depend on a parameter μ which belongs to an interval. Let $\tau(\mu)$ be a stable periodic point of the type n/m of the map $T(\tau, \mu)$. So, the function $G(\tau, \mu) = T^m(\tau, \mu) - 2\pi n$ satisfies the conditions

$$G(\tau(\mu), \mu) - \tau(\mu) = 0 \qquad -1 < G'_\tau(\tau(\mu), \mu) < 1 \quad (4.72)$$

for each τ from the interval. We will discuss what happens with the stable periodic point $\tau(\mu)$ when the derivative $G'_\tau(\tau(\mu), \mu)$ goes over the value $+1$ or -1 as the parameter μ changes its value.

4.6.1 *Saddle-node bifurcation*

Let $\tau_0 = \tau(\mu_0)$ be a neutral fixed point of the map G with the multiplier $d = G'_\tau(\tau_0, \mu_0) = 1$. Assume that $G'_\mu(\tau_0, \mu_0) > 0$, i.e. $G(\tau_0, \mu)$ increases as a function of μ in a neighborhood of μ_0. In Fig.4.21 the graphs of $G(\tau, \mu)$ are shown for three values $\mu_- < \mu_0 < \mu_+$ of the parameter μ, which are close to each other. Fig.4.21 also shows how the fixed points $\tau(\mu)$ of the map G depend on the parameter μ.

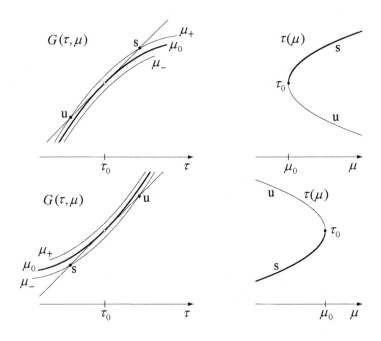

Fig. 4.21 Saddle-node bifurcation. The graphs of the function $G(\tau, \mu)$ and its fixed points $\tau(\mu)$.

Such a bifurcation is called the *saddle-node bifurcation* (see Refs. [50], [51]). In one-sided neighborhood of the value μ_0 of the parameter μ two fixed points exist: a stable and an unstable one. Both tend to one neutral point as $\mu \to \mu_0$ and disappear after crossing over the value μ_0.

The values of fixed points can be locally determined by the derivatives of G at the point (τ_0, μ_0). By Taylor's formula we get

$$G(\tau, \mu) - \tau = \lambda_{01}(\mu - \mu_0) + \frac{\lambda_{20}}{2}(\tau - \tau_0)^2 + \lambda_{11}(\mu - \mu_0)(\tau - \tau_0) + \frac{\lambda_{02}}{2}(\mu - \mu_0)^2 + \dots$$

where

$$\lambda_{ij} = \frac{\partial^{ij} G}{\partial \tau^i \partial \mu^j}(\tau_0, \mu_0). \tag{4.73}$$

We have $\lambda_{01} > 0$. If $\lambda_{20} \neq 0$ then for small values of $|\mu - \mu_0|$ the solutions of the equation $G(\tau, \mu) = \tau$ take the form

$$\tau(\mu) = \tau_0 \pm \sqrt{-\frac{2\lambda_{01}}{\lambda_{20}}(\mu - \mu_0)} + O(|\mu - \mu_0|). \tag{4.74}$$

So, there exists a pair of solutions for μ from a one-sided neighborhood of the point μ_0. One solution is stable and the second is unstable because the derivatives evaluated for these solutions

$$G_\tau'(\tau(\mu), \mu) = 1 + \lambda_{20}(\tau(\mu) - \tau_0) + \ldots = 1 \pm \sqrt{-2\lambda_{01}\lambda_{20}(\mu - \mu_0)} + O(|\mu - \mu_0|)$$

are smaller and greater then one respectively.

4.6.2 Period doubling bifurcation

Let $\tau_0 = \tau(\mu_0)$ be a neutral fixed point of the map G with the multiplier $d = G_\tau'(\tau_0, \mu_0) = -1$. Assume that a stable fixed point $\tau(\mu)$ of the map G exists in a neighborhood of μ_0 for $\mu < \mu_0$. In Fig.4.22 the graphs of $G(\tau, \mu)$ and $G^2(\tau, \mu)$ are shown for three values $\mu_- < \mu_0 < \mu_+$ of the parameter μ. Fig.4.22 also shows how the fixed points of the map G^2 depend on the parameter μ.

Such a bifurcation is called a *period doubling bifurcation* (see Refs. [50], [51]). There are two possibilities.

1. For $\mu < \mu_0$ there exists an isolated stable fixed point of the map G. After crossing over the bifurcation value μ_0, the fixed point loses its stability and in its neighborhood there arises a stable orbit of the map G of period 2 (Fig.4.22a).

2. For $\mu < \mu_0$ there exist a stable fixed point and an unstable orbit of period 2 of the map G, both in a neighborhood of the point τ_0. If $\mu \to \mu_0$, then the fixed point and both points of the orbit tend to τ_0 and there arises one unstable fixed point $\tau(\mu)$ of the map G for $\mu > \mu_0$ (Fig.4.22b).

The values of fixed and periodic points depend on derivatives of the map G at the point (τ_0, μ_0). By Taylor's formula, we get

$$G^2(\tau, \mu) - \tau = -(\lambda_{20}\lambda_{01} + 2\lambda_{11})(\mu - \mu_0)(\tau - \tau_0) +$$

$$+\frac{\lambda_{01}}{2}(\lambda_{20}\lambda_{01} + 2\lambda_{11})(\mu - \mu_0)^2 - \frac{2\lambda_{30} + 3\lambda_{20}^2}{6}(\tau - \tau_0)^3 + \ldots$$

The equation $G^2(\tau, \mu) = \tau$ has three solutions in a neighborhood of the point (τ_0, μ_0). The first is the fixed point

$$\tau(\mu) = \tau_0 + \frac{\lambda_{01}}{2}(\mu - \mu_0) + O(|\mu - \mu_0|^2) \tag{4.75}$$

of the map G. The next pair of solutions

$$\tau_{\pm}(\mu) = \tau_0 \pm \sqrt{-\frac{6(\lambda_{20}\lambda_{01} + 2\lambda_{11})}{2\lambda_{30} + 3\lambda_{20}^2}(\mu - \mu_0)} + O(|\mu - \mu_0|) \qquad (4.76)$$

determines the orbit $\{\tau_-(\mu), \tau_+(\mu)\}$. The equalities $G(\tau_-(\mu), \mu) = \tau_+(\mu)$ and $G(\tau_+(\mu), \mu) = \tau_-(\mu)$ are satisfied. This orbit exists for μ belonging to a one-sided neighborhood of the point μ_0 only. The fixed point (4.75) is stable for $\mu < \mu_0$ and unstable for $\mu > \mu_0$, while the orbit $\{\tau_-(\mu), \tau_+(\mu)\}$ has always the opposite stability to that of the fixed point $\tau(\mu)$.

At the bifurcation point we have $G'_\tau(\tau_0, \mu_0) = \lambda_{10} = -1$, and the expression

$$\lambda_{20}\lambda_{01} + 2\lambda_{11} = 2\frac{d}{d\mu}G'_\tau(\tau(\mu), \mu)|_{\mu=\mu_0}$$

is negative.

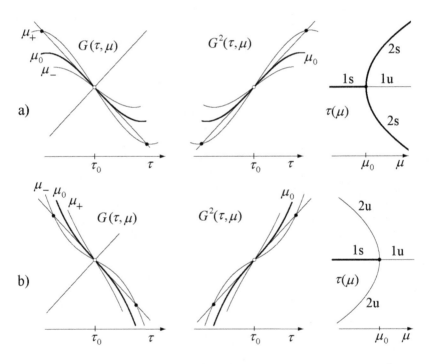

Fig. 4.22 Period doubling bifurcation. The graphs of the functions $G(\tau, \mu)$, $G^2(\tau, \mu)$. The fixed and periodic points $\tau(\mu)$ of the map G: 1s, 1u — fixed points (stable and unstable), 2s, 2u — periodic points of the period 2 (stable and unstable).

The Schwarzian derivative of G at the point τ_0 (for $\mu = \mu_0$) is

$$S_d G(\tau_0, \mu_0) = (\lambda_{10})^{-2}(2\lambda_{30}\lambda_{10} - 3\lambda_{20}^2) = -(2\lambda_{30} + 3\lambda_{20}^2).$$

A fixed point of the map G is the periodic point of the type n/m of the map T. If the function $T(\tau, \mu)$ has negative Schwarzian derivative, then so does $G(\tau, \mu)$. Therefore, for such maps, a stable orbit of the type n/m changes into a stable orbit of the type $2n/2m$ and an unstable orbit of the type n/m as the result of period doubling bifurcation according to (4.75) and (4.76). The period doubling of an unstable orbit shown in Fig.4.22b is not possible for maps with negative Schwarzian derivative.

More complicated bifurcations of stable orbits may occur only in exceptional cases, namely when $\lambda_{10} = 1$ and $\lambda_{01}\lambda_{20} = 0$, or when $\lambda_{10} = -1$ and $(\lambda_{01}\lambda_{20} + +2\lambda_{11})(2\lambda_{30} + 3\lambda_{20}^2) = 0$. We omit discussion of these cases.

4.6.3   The Feigenbaum cascade

Suppose that a parameter changes not only locally in a neighborhood of the bifurcation point but passes through an interval. It so happens that an infinite number of bifurcation points occur, and even their condensation points are not isolated. We will now describe the phenomenon of successive bifurcations of stable orbits for a family of unimodal maps.

Let us consider a trajectory $\{\tau_0, \tau_1, \tau_2, \ldots\}$ of the map (4.8) for parameters (μ, a) belonging to the domain (2a) shown in Fig.4.20. For a given initial point τ_0, the successive terms of the trajectory can be obtained by the simple geometrical construction shown in Fig.4.23.

In Fig.4.24a it is shown how the properties of the trajectory depend on values of the parameter a for a fixed $\mu = -0.12$. For each of 1200 values of a which are uniformly distributed in the interval $[2.1, 3.5]$, there were calculated 600 terms of the trajectory of a random value $\tau_0 \in I_0 = (z, z + 2\pi)$. The end-points of the interval I_0 are the unstable fixed points of T (see Section 4.5.2). The first 200 terms are neglected, and the sets $S = \{\tau_n : n = 200, \ldots, 600\}$ versus a are shown in the figure. This picture practically does not depend on the initial value τ_0.

If $a < a_0 \approx 2.14$ then there exists one stable fixed point which attracts all initial points $\tau_0 \in I_0$. For $a \in (a_0, a_1)$ there exists a stable orbit of the period 2 which attracts all initial points $\tau_0 \in I_0$ except one unstable fixed point. For $a \in (a_1, a_2)$ there exists a stable orbit of the period 4 which attracts all initial points $\tau_0 \in I_0$ except one unstable fixed point and one unstable periodic orbit of the period 2, etc. On the a-axis there

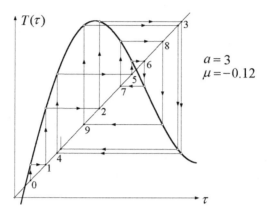

Fig. 4.23   Geometrical construction of the trajectory $\{\tau_0, \tau_1, \tau_2, \ldots\}$ determined by the map (4.8). The point marked by the number k has both coordinates equal to τ_k.

exists an infinite increasing sequence $\{a_0, a_1, a_2, \ldots\}$ of the period doubling bifurcation points, converging to the value $a_\infty \approx 2.836$. At the point a_k a stable orbit of the period 2^k loses its stability and there arises a stable orbit of period 2^{k+1} (and unstable orbit of period 2^k). The rate of convergence of the sequence $\{a_k\}$ is the same as for a geometric progression with the ratio

$$\delta^{-1} = \lim_{k \to \infty} \frac{a_\infty - a_k}{a_\infty - a_{k-1}} = 0.214169377\ldots \qquad (4.77)$$

Such a sequence of period doubling bifurcations of stable orbits is called the *Feigenbaum cascade* (see Refs. [18], [15]).

For $a > a_\infty$ there is a set M of the values of parameter a, for which the map T has no stable orbit. The set M has a positive Lebesgue's measure but does not contain any interval. In particular, the set M contains a point a_k^0 for which the trajectory starting from the critical point hits the unstable orbit of period 2^k. According to the Singer theorem, if $a = a_k^0$ then no stable orbit exist. The set of points a_k^0 forms the decreasing sequence $\{a_0^0, a_1^0, a_2^0, \ldots\}$ converging to a_∞ with the rate determined by the ratio (4.77).

For $a > a_\infty$ there exists an infinite number of such values of parameter a, for which stable orbits of different periods arise as a result of the saddle-node bifurcations (e.g. period 3 for $a \approx 3.19$, period 6 for $a \approx 2.907$, period 12 for $a \approx 2.852$, period 10 for $a \approx 2.881$ etc.). If a increases, then each of such orbits doubles its period an infinite number of times according to the Feigenbaum cascade described above.

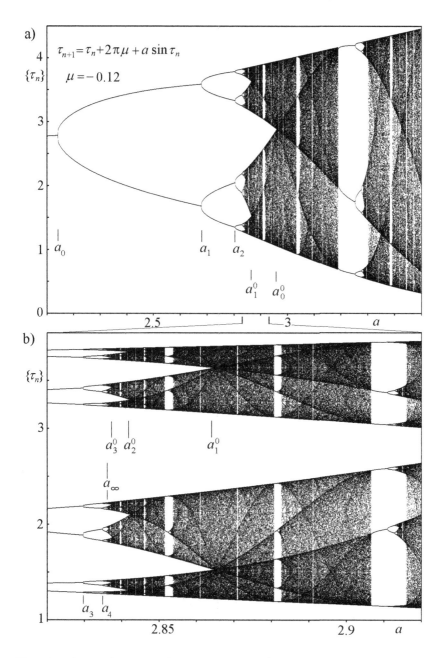

Fig. 4.24 Graph of sets $S = \{\tau_n : n = 200, ..., 600\}$ versus a, for two ranges of the parameter a.

The properties presented above and shown in Fig.4.24 are typical for a large class of one-parameter families of maps the graphs of which are similar to the parabola $f_a(\tau) = a - \tau^2$ with the parameter $a \in [-0.25, 2]$. This class of maps will be described more precisely in Sections 4.6.7 and 4.7.3.

4.6.4   Invariant measures

Let us consider the sequence $\{f^n(\tau_0)\}$, $n = 0, 1, 2, \ldots$ determined by a smooth function f which maps an interval I into itself. Let the initial term τ_0 take all values from the interval I with the probability density function

$$p_0(t) = \lim_{\varepsilon \to 0} \frac{1}{\varepsilon} P(t \le \tau_0 < t + \varepsilon), \quad t \in I. \tag{4.78}$$

The density function $p_0(t)$ has the standard properties: $p_0(t) \ge 0$ and $\int_I p_0(t)dt = 1$.

The density function of the next term $\tau_1 = f(\tau_0)$ is determined by the Perron-Frobenius operation $\mathcal{P}$ (see Refs. [52], [15]), i.e.

$$p_1(t) = (\mathcal{P}p_0)(t) = \sum_i \frac{1}{|f'(\alpha_i)|} p_0(\alpha_i), \quad \text{where } f(\alpha_i) = t. \tag{4.79}$$

A geometrical interpretation of the operation $\mathcal{P}$ is shown in Fig.4.25. For example,

$$\text{if } f(\tau) = a - \tau^2, \quad \text{then } (\mathcal{P}p)(t) = \frac{p\left(\sqrt{a-t}\right) + p\left(-\sqrt{a-t}\right)}{2\sqrt{a-t}}.$$

The density functions

$$p_2(t), \ p_3(t), \ p_4(t), \ \ldots, p_n(t), \ \ldots$$

of the next terms of the sequence $\{f^n(\tau_0)\}$ are obtained in the same manner: $p_n = \mathcal{P}(p_{n-1})$. If the sequence of the functions $\{p_n(t)\}$ is convergent then the limit function $p_\infty(t)$ fulfils the equation $p_\infty = \mathcal{P}(p_\infty)$. A fixed point of the operation $\mathcal{P}$ is called a *density of the invariant measure* of the map f. A closure of the set $\{t : p_\infty(t) \ne 0\}$ is called the *support* of the invariant measure. If the map f has a stable periodic orbit $\{\tau_1, \tau_2, \ldots, \tau_m\}$ which attracts almost all points of the interval I, then the invariant measure is discrete. The support consists of the points of the stable orbit, and

$$p_\infty(t) = \frac{1}{m} \sum_{i=1}^{m} \delta(t - \tau_i),$$

where $\delta(t)$ is Dirac's delta function. However, there exist maps f for which the function $p_\infty(t)$ is the derivative of an absolutely continuous function

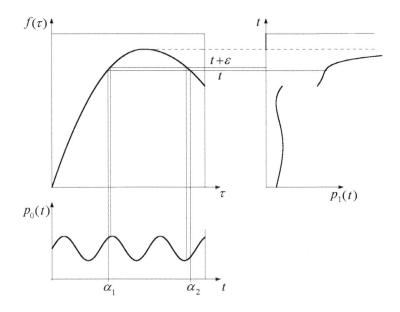

Fig. 4.25   Illustration to the definition of the Perron-Frobenius operation $p_1 = \mathcal{P}(p_0)$.

(the invariant measure is absolutely continuous), and for which the support consists of intervals.

An invariant measure is called *ergodic* (see Refs. [57], [60]) if for each continuous function ψ (defined on I) and for almost each trajectory $\{\tau_0, \tau_1, \tau_2, \ldots\}$, where $\tau_k = f(\tau_{k-1})$, the mean value of the function ψ calculated along the trajectory is equal to the expected value calculated with respect to the invariant measure

$$\lim_{N \to \infty} \frac{1}{N} \sum_{k=0}^{N-1} \psi(\tau_k) = \int_I \psi(t) p_\infty(t) dt. \tag{4.80}$$

The set of values τ_0 for which the equality (4.80) is not true has the invariant measure equal to zero, i.e. the equality (4.80) holds with probability equal to one.

Let us now return to the map (4.8) with the parameters μ, a from the domains (2a) or (2b) shown in Fig.4.20. In the μa-plane there exists an uncountable set of points for which the map T has an absolutely continuous and ergodic invariant measure. In particular, there exists a family of curves $C_k = \{\mu, a_k^0(\mu)\}$, $k = 0, 1, 2 \ldots$, where the decreasing sequence $\{a_k^0(\mu)\}$

converges to a limit function $a_\infty(\mu)$. If $(\mu, a) \in C_k$ then the support of the invariant measure of the map T consists of 2^k intervals. The values a_0^0, a_1^0, a_2^0, a_3^0 and a_∞ for $\mu = -0.12$ are marked in Fig.4.24.

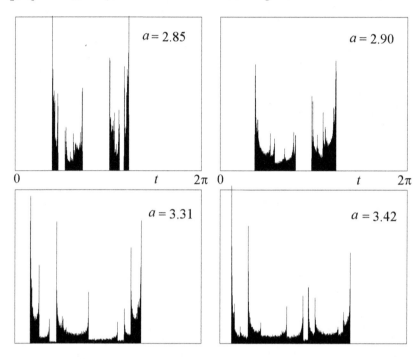

Fig. 4.26   Examples of densities of the invariant measures of the map (4.8) for $\mu = -0.12$ and for the four values of a.

An invariant measure (absolutely continuous and ergodic) can be determined numerically. Let us consider the equality (4.80) with $\psi(t) = \frac{1}{\varepsilon}$ for $t \in [\tau, \tau + \varepsilon)$ and $\psi(t) = 0$ for $t \notin [\tau, \tau + \varepsilon)$. Then, for a fixed small value $\varepsilon > 0$ and for a sufficiently large N we get the approximate equality

$$p_\infty(\tau) \approx \frac{1}{\varepsilon} \int_\tau^{\tau+\varepsilon} p_\infty(t) dt \approx \frac{n(\tau, \varepsilon, N)}{\varepsilon N}, \qquad (4.81)$$

where $n(\tau, \varepsilon, N)$ is the number of terms of the truncated sequence $\{\tau_1, ..., \tau_N\}$ which belong to the interval $[\tau, \tau + \varepsilon)$. In Fig.4.24 the more intensive the shading, the larger the density of the invariant measure. In Fig.4.26 examples of densities of the invariant measures p_∞ of the map (4.8) are shown. They are calculated by formula (4.81) for $\varepsilon = 2\pi/600$, $N = 60000$ and for the four values of (μ, a).

4.6.5 The Liapunov exponent

Although all terms of the sequence $\{f^n(\tau_0)\}$, $n = 0, 1, 2, 3, \ldots$ are completely determined by the initial value τ_0, a sufficiently distant term τ_N can be so sensitive to a change of τ_0 that, practically, we can not specify its value. If δ_0 is a sufficiently small increment of τ_0, then

$$\tau_{k+1} + \delta_{k+1} = f(\tau_k + \delta_k) \approx f(\tau_k) + f'(\tau_k)\delta_k, \quad k = 0, 1, 2, 3, \ldots$$

and, consequently, $\delta_{k+1} \approx f'(\tau_k)\delta_k = \prod_{i=0}^{k} f'(\tau_i)\delta_0$. For a fixed integer N, if δ_0 is sufficiently small, then we have

$$|\delta_N| \approx e^{N\lambda_N(\tau_0)}|\delta_0|,$$

where

$$\lambda_N(\tau_0) = \frac{1}{N} \ln \left| \frac{df^N}{d\tau}(\tau_0) \right| = \frac{1}{N} \sum_{k=0}^{N-1} \ln |f'(\tau_k)|. \tag{4.82}$$

If the limit

$$\lambda(\tau_0) = \lim_{N \to \infty} \lambda_N(\tau_0) \tag{4.83}$$

exists, then it is called the *Liapunov exponent of the trajectory* $\{f^n(\tau_0)\}$ (see Refs. [52], [38]). A periodic trajectory $\{f^n(\tau_0)\}$ of the period m with the multiplier d has the Liapunov exponent $\lambda(\tau_0) = \frac{1}{m} \ln |d|$. This trajectory is stable for $\lambda(\tau_0) < 0$, superstable for $\lambda(\tau_0) = -\infty$, unstable for $\lambda(\tau_0) > 0$ and neutral for $\lambda(\tau_0) = 0$.

If the limit (4.83) is independent of τ_0 for almost every τ_0, then it is called the *Liapunov exponent of the map* f. For example, if $\{\tau_0, \tau_1, \ldots, \tau_{m-1}\} \bmod 2\pi$ is a stable periodic orbit of the map f with a multiplier $d \in (-1, 1)$ then the Liapunov exponent $\lambda(\tau)$ does not depend on τ for all $\tau \in A_{\mathrm{orb}}(\tau_0)$. So, $\lambda = \frac{1}{m} \ln |d|$ is the Liapunov exponent of the map f which is defined on the domain restricted to the attractive set $A_{\mathrm{orb}}(\tau_0)$ of the stable orbit.

If a map f has a positive Liapunov's exponent then almost all trajectories are repelling, and sequences $\{f^n(\tau)\}$ are really sensitive to a change of initial points τ_0 (see Refs. [21], [16]).

If the map f has an ergodic invariant measure, then there exists a relation between this measure and the Liapunov exponent. Indeed, replacing $\psi(t)$ by $\ln |f'(t)|$ in (4.80), we get

$$\lambda = \int_I \ln |f'(t)| p_\infty(t) dt. \tag{4.84}$$

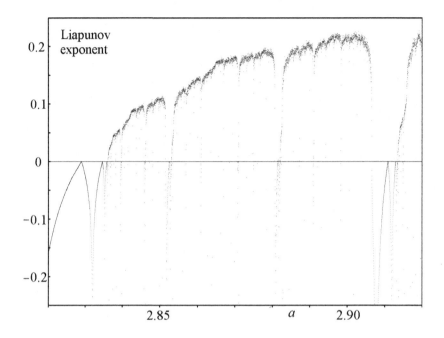

Fig. 4.27   The Liapunov exponent versus the parameter a, for the map (4.8) and for $\mu = -0.12$.

The map f which is defined on the domain restricted to the support of the ergodic invariant measure with density $p_\infty(t)$, has the Liapunov exponent (4.84).

Example 4.5. Let us assume that the Liapunov exponent of the map T is equal $\lambda = 0.2$, just like in Fig.4.27 for $\mu = -0.12$ and $a \approx 2.9$. The initial value τ_0 of the sequence $\{T^n(\tau_0)\}$ is given with the error $|\delta_0| = 10^{-6}$. So, the term $T^N(\tau_0)$ of the number $N = 76$ is calculated with the error $|\delta_N| \approx |\delta_0| e^{\lambda N} = 3.99...$ which exceeds the length of the interval containing all terms of the sequence.

In Fig.4.27 the Liapunov exponent λ of the map (4.8) is shown versus the parameter a for the same (μ, a) as in Fig.4.24b. It is such a domain of parameters (the domain 2a shown in Fig.4.20) where all trajectories are attracted to the interval I_L which contains only one critical point c_1.

The map (4.8) with two critical points c_1 and c_2 has two Liapunov's exponents λ_1, λ_2 which are independent of τ_0 for almost every τ_0 from small neighborhoods of critical points c_1 and c_2 respectively. A numerical exper-

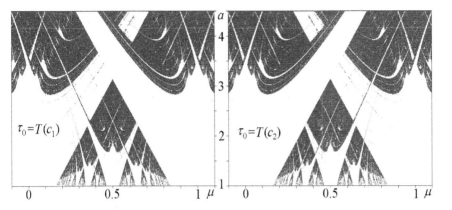

Fig. 4.28   Loci of parameters (μ, a) of the map (4.8) with the positive Liapunov exponents λ_1, λ_2 of two trajectories which start from the critical values $T(c_1)$ and $T(c_2)$.

iment is shown in Fig.4.28. For each point (μ, a), where $\mu \in (-0.1, 1.1)$, $a \in (1, 4.5)$, the Liapunov exponents λ_1, λ_2 of trajectories $\{T^n(\tau_{01})\}$, $\{T^n(\tau_{02})\}$ of two critical values $\tau_{01} = T(c_1)$ and $\tau_{02} = T(c_2)$ are calculated. If the Liapunov exponent takes a positive value then the point (μ, a) is black in the picture. If it takes negative value then the point (μ, a) is white. There exist many points (μ, a) for which these two Liapunov's exponents have opposite signs.

4.6.6   Skeleton of superstable orbits

Let $\{\tau_0, \ldots, \tau_{m-1}\}(mod\, 2\pi)$ be a periodic orbit of the type n/m of the map $T(\tau; \mu, a)$ which depends on the parameters (μ, a). Let the multiplier

$$d(\mu, a) = \prod_{i=0}^{m-1} T'(\tau_i; \mu, a)$$

be a continuous function of (μ, a). The orbit is stable for $d(\mu, a) \in (-1, 1)$ and superstable for $d(\mu, a) = 0$. A superstable orbit is easier to analyze than other stable orbits because it contains a critical point and attracts the neighboring trajectories faster. If the map $T(\tau; \mu, a)$ has a superstable orbit for (μ_*, a_*), then it has a stable orbit of the same type for each (μ, a) belonging to a neighborhood of (μ_*, a_*). So, the superstable orbits are worth special attention.

Some results of the MacKay and Tresser theory of a two-parameter

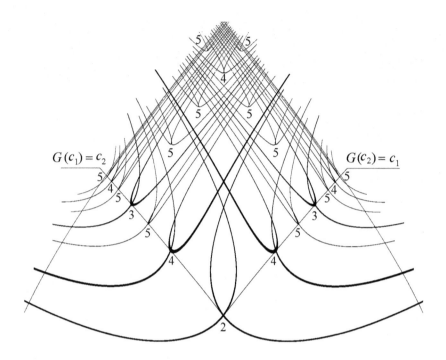

Fig. 4.29   Bones of superstable periodic orbits of the periods 2, 3, 4, 5 for a two-
-parameter family of bimodal maps (4.69) for (μ, a) belonging to the domains (1),
(2a), (2b) shown in Fig.4.20.

family of smooth bimodal maps will be presented below (see Refs. [36], [37]).

Let the map $G(\tau; \mu, a)$ depend on parameters μ, a and satisfy the following conditions:

(1) G has two unstable fixed points τ_1, τ_2 and maps the interval $[\tau_1, \tau_2]$
onto itself,

(2) G has two critical points $c_1 < c_2$, and in the intervals $L = (\tau_1, c_1)$,
$C = (c_1, c_2)$, $R = (c_2, \tau_2)$ the derivative G' is positive, negative and positive
respectively,

(3) G has negative Schwarzian derivative,

(4) the parameter a changes the variation of the function G, and the
parameter μ shifts the graph of G up or down, as in (4.69).

For each fixed $k \geq 2$ the set of all superstable orbits $\{\tau_1, \ldots, \tau_k\}$ of
the period k decomposes into some classes, but we do not describe this
decomposition. A number of these classes is no greater than $(k - 1)!$. To
each class there is attached a so called *bone*, i.e. the locus of points (μ, a) for

which there exists a superstable orbit of this class. In Fig.4.29 the bones of superstable orbits of the periods 2,3,4,5 are shown for the map (4.69). The range of the parameters μ, a is the same as in the domains (1), (2a), (2b) in Fig.4.20. The set of all bones, for all superstable orbits of all periods greater then one, forms the so called *skeleton*. The family of the maps $G(\tau; \mu, a)$ has exactly one bone for orbits of period 2, two bones for orbits of period 3, five bones for orbits of period 4, twelve bones for orbits of period 5, etc.

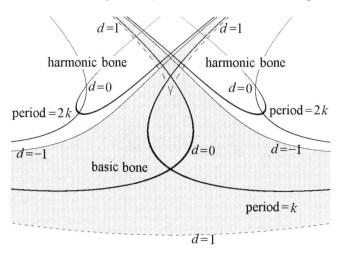

Fig. 4.30   Three bones of superstable periodic orbits (the basic bone for orbits of period k, and two harmonic bones for orbits of period $2k$). The domain with stable periodic orbits of period k is marked in grey.

Each bone has a specific shape, similar to two parabolas which have two intersection points. In the lower intersection point there exists *doubly superstable orbit*, i.e. the orbit which contains both critical points. An example of three bones is shown in Fig.4.30. In a neighborhood of each bone there exists a domain of parameters μ, a (shaded in Fig.4.30) for which $G(\tau; \mu, a)$ has a stable periodic orbit $\{\tau_1, \ldots, \tau_k\}$ with a multiplier $d \in$ $\in (-1, +1)$. The boundaries of this domain consist of two dashed lines (lower and upper) with the multiplier $d = 1$ and two solid lines with the multiplier $d = -1$.

If a point (μ, a) crosses over from below one of the solid lines, then the so called *harmonic orbit* $\{\tau'_1, \ldots, \tau'_k, \tau'_{k+1}, \ldots, \tau'_{2k}\}$ of the period $2k$ (with the multiplier $d = 1$) appears as a result of the period doubling bifurcation. The location of two *harmonic bones* (for superstable orbits of the period $2k$) with

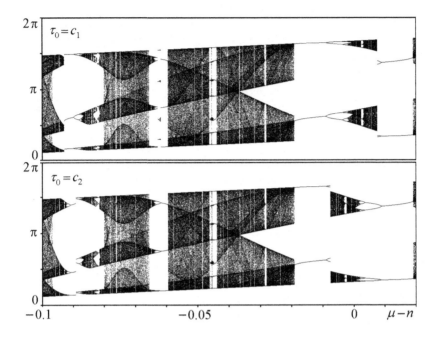

Fig. 4.31   The sets $\{G^n(c_i) \,(mod\,2\pi) : n = 200, ..., 600\}$ versus $(\mu - n)$ for two critical
points c_1 and c_2 of the map (4.69) with $a = 3.55$.

respect to the *basic bone* (for superstable orbits of the period k) is shown
in Fig.4.30 above the lines of the period doubling bifurcations. If a point
(μ, a) crosses over from above the upper dashed line, then a stable orbit of
the period k disappears as a result of the saddle-node bifurcation. If a point
(μ, a) crosses over from above the lower dashed line, then a stable orbit of
the period k either disappears as a result of the saddle-node bifurcation
(provided that it is not a harmonic orbit of another basic orbit) or changes
into a stable orbit of the period $\frac{k}{2}$ with $d = -1$ as a result of the period
doubling bifurcation taken in the inverse direction.

In μa-plane there exists an infinite number of such subregions where the
map $G(\tau; \mu, a)$ has two stable orbits. There also exists an infinite number
of points where $G(\tau; \mu, a)$ has two superstable orbits.

The results of a numerical experiment are shown in Fig.4.31 for the map
(4.69) with $a = 3.55$. The steady-state trajectories $\{\tau_0, \tau_1, \tau_2, \dots\}\,(mod\,2\pi)$
of two critical points c_1 and c_2 are calculated. For 1200 values of the
parameter $(\mu-n)$ which are uniformly distributed in the interval $[-0.1, 0.02]$,
the sets $\{G^n(c_i)\,(mod\,2\pi) : n = 200, ..., 600\}$ are shown for $i = 1, 2$.

For example, if $(\mu-n)=0.005$ then two different steady-states exist: the periodic orbit of the period 4 (for $\tau_0 = c_2$), and *chaotic* trajectories with the positive Liapunov exponent (for $\tau_0 = c_1$). However, for many values of the parameter $(\mu-n)$ there exists exactly one steady-state which attracts the trajectories of both critical points.

4.6.7 *The Feigenbaum cascade (continuation)*

The Feigenbaum cascade is a most spectacular phenomenon observed in the phase locked loop and in many other dynamical systems in nature. Let the output signal of DPLL be periodic of the period $\frac{2\pi}{\omega}$. Its spectrum has a basic component of the frequency ω and harmonic components. If the change of a parameter causes successive doubling of the output signal period, then in the spectrum there appear successively components of frequencies $\frac{\omega}{2}, \frac{\omega}{4}, \frac{\omega}{8}, \frac{\omega}{16}, \ldots$, and their harmonics. For the limit value of the parameter (e.g. for a_∞ introduced in Section 4.6.3) the spectrum of the output signal happens to be dense in a neighborhood of zero of the frequency axis. In the place of the signal with a discrete spectrum there appears a noise.

Similarly, the Feigenbaum cascade can be detected using an oscilloscope. In Fig.4.32 the steady-state *input-output relations* are shown for DPLL described by the map (4.8) for $\mu = 0.88$ and four values of the parameter a. More precisely, the figure shows lines which parametric representations take the form

$$x = U_{\text{inp}}(\omega t) = \sin \omega t$$
$$y = U_{\text{out}}(\theta(t)) = \cos\left(2\pi \frac{\omega t - \tau_n}{\tau_{n+1} - \tau_n}\right), \quad \text{for } \omega t \in (\tau_n, \tau_{n+1}),$$

where $n = 701, ..., 900$. The initial value τ_0 is a random one. The types of stable periodic oscillations are successively: $1/1$, $2/2$, $4/4$. If a increases, then the next types, i.e. $8/8$, $16/16, \ldots$ can be observed. For $a > a_\infty$ (the last picture) the stable oscillations can be non-periodic (*chaotic*). So, the Feigenbaum cascade is easy to detect experimentally.

Now we will present a short sketch of an extremely ingenious theory proposed by Feigenbaum to explain the phenomenon of the period doubling bifurcation. Its understanding demands however some imagination related to simple geometrical objects in infinite dimensional spaces.

Let $\mathcal{U}$ denote the space of *unimodal functions* satisfying the following conditions:

1) $f(0) = 1$, $f'(0) = 0$, $f''(0) < 0$,

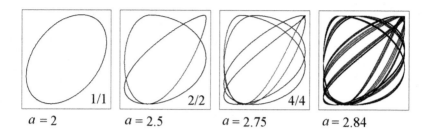

$a = 2$ $a = 2.5$ $a = 2.75$ $a = 2.84$

Fig. 4.32   The input-output relations of DPLL for the oscillations of the types 1/1, 2/2, 4/4 and a "chaotic" one.

2) f maps the interval $[f(1), 1]$ into itself, $f'(x) > 0$ for $x \in [f(1), 0)$ and $f'(x) < 0$ for $x \in (0, 1]$,

3) f has negative Schwarzian derivative.

We define the so called *doubling operation* $\mathcal{D}$ on the space $\mathcal{U}$. The notation $g = \mathcal{D}(f)$ means that

$$g(x) = \frac{1}{\lambda} f[f(\lambda x)], \quad \text{where } \lambda = f(1) < 0. \tag{4.85}$$

This operation consists of a twofold composition of the function f (Fig. 4.33) and stretching both axes scales with the inversion of their directions ($\lambda < 0$). The function g satisfies the conditions 1) and 3) presented above. The operation $\mathcal{D}$ is suitable for analysis of superstable orbits. The map f has a superstable orbit of the period $2k$ if and only if the map $g = \mathcal{D}(f)$ has a superstable orbit of the period k.

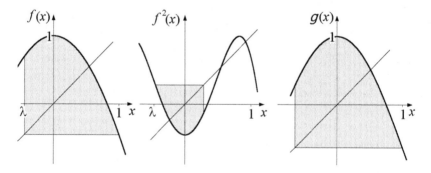

Fig. 4.33   Graphs of the functions $f(x)$, $f^2(x)$ and $g(x)$.

The operation $\mathcal{D}$ has a fixed point $\tilde{f} = \mathcal{D}(\tilde{f})$ in the space $\mathcal{U}$, i.e. there exists the function

$$\tilde{f}(x) \approx 1 - 1.52763x^2 + 0.104815x^4 - 0.0267057x^6 + \dots$$

which satisfies the functional equation $\lambda f(x) = f[f(\lambda x)]$, with $\lambda = \tilde{f}(1) \approx$ $\approx -0.3995...$

In a small neighborhood of $\tilde{f}$ properties of the operation $\mathcal{D}$ are determined by the Frechet derivative $H = \mathcal{D}'(\tilde{f})$ of the operation $\mathcal{D}$ at the point $\tilde{f}$. The linear operation H has a one-dimensional invariant subspace with the eigenvalue $\delta = 4.66920...$ The rest of the spectrum belongs to the interior of the unit disc. So, in a neighborhood of the point $\tilde{f} \in \mathcal{U}$ there exists a one-dimensional repellent invariant manifold W_u tangent at the point $\tilde{f}$ to the invariant subspace of the operation H corresponding to the eigenvalue δ. The manifold W_u crosses at the point $\tilde{f}$ the stable invariant manifold W_s of codimension one. The action scheme for the operation $\mathcal{D}$ in a neighborhood of $\tilde{f}$ of the space $\mathcal{U}$ is shown in Fig.4.34.

Let

$$\Sigma_1 = \{f : f \in \mathcal{U}, f^2(0) = 0\}$$

denote the set of the functions (maps) f which have a superstable orbit of the period 2. The manifold W_u intersects transversally the *hypersurface* Σ_1. One can now form the sequence of hypersurfaces

$$\Sigma_k = \{f : f \in \mathcal{U}, \mathcal{D}(f) \in \Sigma_{k-1}\}, \quad k = 2, 3, 4, \dots$$

coming closer to the stable manifold W_s. The hypersurface Σ_k is a set of such unimodal maps f which have a superstable orbit of the period 2^k. The manifold W_s contains the maps f which have an attracting orbit of the "period 2^∞".

Now let us consider a one-parameter family of functions $f(\cdot, \mu)$ close to the function $\tilde{f}(\cdot)$. A geometrical image of such a family is the line f_μ placed in the space $\mathcal{U}$ (Fig. 4.34). Let it intersect transversally the manifold W_s and the hyper-surfaces Σ_k for successive values $\mu_1, \mu_2, \mu_3, \dots$ of the parameter μ. So, the map $f(\cdot, \mu_k) \in \Sigma_k$ has a superstable orbit of the period 2^k. Let μ_∞ denote such a value of the parameter for which $f(\cdot, \mu_\infty) \in$ $\in W_s$. The sequence $\{\mu_k\}$ converges to μ_∞ as a geometric progression with the ratio δ^{-1} which is determined by the eigenvalue of the operation H, i.e.

$$\lim_{k \to \infty} (\mu_\infty - \mu_k)\delta^k = const. \neq 0.$$

So, the Feigenbaum number $\delta = 4.66920...$ is a universal constant common to the whole set of unimodal functions $f(x, \mu)$ which depend on a parameter μ. The second universal constant is the number $-\tilde{f}(1) = 0.3995...$ It

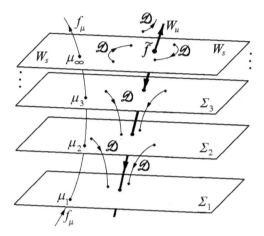

Fig. 4.34   Action scheme for the doubling operation $\mathcal{D}$ in the space $\mathcal{U}$ of unimodal functions.

determines the rate of condensation of the points of the superstable orbits with doubled periods in a neighborhood of the critical point $x = 0$.

The proofs of the geometrical facts presented above are given in Ref. [15].

4.7   Bifurcation of the rotation interval

Let us consider the bimodal map $T(\tau; \mu, a)$ given by (4.8) or close to it. Let D_N denote a region of the parameters μ, a for which there exists a stable periodic point of the type $N/1$ with an immediate attractive interval of the length equal to 2π. In this region the DPLL system divides an input signal frequency by N, independently of an initial value, and preserves the constant phase difference. The border of the region D_N consists of three kinds of lines (Fig.4.35):

  snb-lines of saddle-node bifurcation,
  pdb-line of period doubling bifurcation,
  rib-lines of rotation interval bifurcation.

The first and the second kinds of bifurcations were discussed in Section 4.6. We present now the bifurcation "from rotation number to rotation interval".

If $(\mu, a) \in D_N$ then the map $T(\tau) - 2\pi N$ has an unstable fixed point

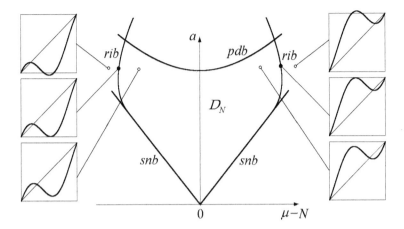

Fig. 4.35   Borders snb, pdb, rib of the region D_N of the map (4.8) and diagrams of the map (4.86) for six selected values of parameters which are marked in a neighborhood of the rib-lines.

z and maps the interval $(z, z + 2\pi)$ onto itself. This whole interval is an attractive domain of a stable fixed point $z + \tau_s$. It is convenient to translate the point $\big(z, T(z)\big)$ to the origin of coordinates by introducing the new map

$$G(\tau) = T(z + \tau) - z - 2\pi N. \qquad (4.86)$$

Evidently, the map $G(\tau)$ has the unstable fixed point $\tau = 0$ and a stable point $\tau = \tau_s$. Since $T(\tau)$ is bimodal, then $G(\tau)$ has two critical points $c < c'$ and $0 < G(c') < G(c) < 2\pi$. The rotation number of the map G is equal to zero.

If (μ, a) lies on the right branch of the rib-line then $G(c) = 2\pi$ and there exists a *heteroclinic semi-trajectory* $\{G^n(c)\}$, $n = ..., -2, -1, 0, 1$ from the unstable fixed point zero to the unstable fixed point 2π. All points of the interval $(0, 2\pi)$ excluding this semi-trajectory are attracted to τ_s. If the point (μ, a) lies outside the region D_N after crossing the right branch of rib-line, then $G(c) > 2\pi$ as in Fig.4.36. It follows that $G_+(\tau) > \tau$ for each τ, and the map (4.86) has a rotation interval $[0, \rho(G_+)]$.

Similarly, if (μ, a) lies on the left branch of the rib-line then $G(c') = 0$, and there exists a *heteroclinic semi-trajectory* $\{G^n(c')\}$, $n = ... - 2, -1, 0, 1$ from the unstable fixed point 2π to the unstable fixed point zero. All points of the interval $(0, 2\pi)$ excluding this semi-trajectory are attracted to τ_s. If the point (μ, a) lies outside the region D_N after crossing the left branch of *rib*-line, then $G(c') < 0$. It follows that $G_-(\tau) < \tau$ for each τ, and the map (4.86) has a rotation interval $[\rho(G_-), 0]$.

Since both bifurcations of rotation intervals can be analyzed exactly in the same manner, let us consider the first one only.

4.7.1   A simplified mapping

For more transparency, we assume a simpler formula of the bimodal function $G(\tau)$.

$1°$ In a small neighborhood of the local maximum point c there is

$$G(\tau) = 2\pi + \varepsilon - b(\tau - c)^2 \quad \text{for} \ \tau \in I_0 = \left(c - \sqrt{\varepsilon/b}, \ c + \sqrt{\varepsilon/b}\right). \quad (4.87)$$

The number $\varepsilon = G(c) - 2\pi$ is a new positive parameter which replaces μ and a. The value of the parameter $b = -\frac{1}{2}G''(c) > 0$ is fixed. Evidently, $G(\tau) > 2\pi$ for $\tau \in I_0$.

$2°$ In the right-hand side neighborhood of zero there is

$$G(\tau) = q\tau \quad \text{for} \ q\tau \in \left(0, \ c + \sqrt{\varepsilon/b}\right), \quad (4.88)$$

where $q = G'(0)$ is a large number, greater than $\dfrac{c + \sqrt{\varepsilon/b}}{c - \sqrt{\varepsilon/b}}$.

$3°$ The interval $A_0 = \left(c + \sqrt{\varepsilon/b}, \ 2\pi\right)$ is the immediate attractive interval of the stable fixed point τ_s.

The graph of the function $G(\tau)$ in the respective intervals is similar to the graphs of functions (4.8) and (4.7) for parameters which belong to a neighborhood of the right branch of the *rib*-line. For example, the graph shown in Fig.4.36 satisfies approximately the conditions $1°, 2°, 3°$ with the parameters $q = 2.82$, $b = 1.12$, $c = 2.72$, $\varepsilon = 0.308$.

Let us decompose the interval $(0, 2\pi)$ into three sets A, I and H. The set A is the attractive set of the point τ_s. It contains an infinite number of disjoint intervals $A_0, A_1, A_2, \ldots$, where A_0 is the immediate attractive interval and $G(A_n) = A_{n-1}$ for $n = 1, 2, 3, \ldots$ The set I contains $I_0 = \{\tau : G(\tau) > 2\pi\}$ and its preimages $I_1, I_2, I_3, \ldots$, where $G(I_n) = I_{n-1}$ for $n = 1, 2, 3, \ldots$ The set H is the common border of the sets A and I, and it consists of two heteroclinic semi-trajectories from the unstable fixed point zero to the unstable fixed point 2π. We have

$$(0, 2\pi) = A \cup I \cup H, \quad \text{where} \ A = \bigcup_{k=0}^{\infty} A_k, \ I = \bigcup_{k=0}^{\infty} I_k. \quad (4.89)$$

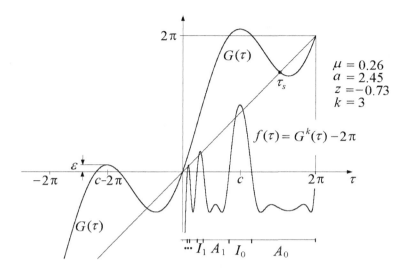

Fig. 4.36 Graphs of functions $G(\tau)=\tau+2\pi\mu+a\sin(\tau+z)$ and $f(\tau)=G^k(\tau)-2\pi$.

The measure of the set I is

$$\mathrm{mes}\,I = \sum_{n=0}^{\infty} \mathrm{mes}\,I_n = \frac{2}{1-q^{-1}}\sqrt{\frac{\varepsilon}{b}}. \qquad (4.90)$$

Moreover, $\mathrm{mes}\,H = 0$ and $\mathrm{mes}\,A = 2\pi - \mathrm{mes}\,I$.

The function $G(\tau)$ depends on the parameters c, b, q, ε but bifurcation of the rotation interval depends mainly on ε. Therefore, two notations $G(\tau)$ or $G(\tau;\varepsilon)$ will be used.

4.7.2 Superstable periodic orbits of the type $1/k$

For a given small value $\varepsilon > 0$, the segment of the trajectory $\{G^n(c-2\pi;\varepsilon)\}$ of the critical point $c-2\pi$ takes the form

$$c-2\pi,\ \varepsilon,\ q\varepsilon,\ q^2\varepsilon,\ q^3\varepsilon,\ \ldots,\ q^{k-1}\varepsilon \qquad (4.91)$$

as long as $q^{k-1}\varepsilon < c + \sqrt{\varepsilon/b}$. It follows from (4.87)–(4.88) and from the identity $G(\tau + 2\pi) = G(\tau) + 2\pi$. If $q^{k-1}\varepsilon = c$, then $G^k(c;\varepsilon) = c + 2\pi$, and (4.91) is a superstable periodic orbit of the type $1/k$. Moreover, $1/k$ is equal to the rotation number of the map $G_+(\tau)$ and is the length of the rotation interval of both maps $G(\tau)$ and $T(\tau)$.

Let $\varepsilon_{ss}(k) = cq^{1-k}$ denote such value of the parameter ε for which superstable orbit of the type $1/k$ exists. For small values of ε the length of

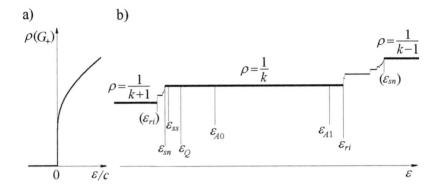

Fig. 4.37   Graph of the length of the rotation interval (4.92) versus the parameter $\varepsilon = G(c) - 2\pi$ and an enlarged segment of the graph.

the rotation interval is

$$\rho(G_+) = -\frac{1}{\log_q(\varepsilon/cq)}, \quad \text{for } \varepsilon = \varepsilon_{ss}(k). \tag{4.92}$$

The rotation interval expands very quickly. In the origin of coordinates, the graph of $\rho(G_+)$ versus ε has the tangency of an infinite order to the vertical line (see Fig.4.37a and Ref. [30]). If ε increases, e.g. from $\varepsilon_{ss}(k)$ to $\varepsilon_{ss}(k{-}1)$, then $\rho(G_+)$ is the Cantor step-function of the parameter ε (Fig.4.37b) and it increases from $1/k$ to $1/(k-1)$. If G has negative Schwarzian derivative, then at most two stable periodic orbits exist. One of them is the stable fixed point $\tau_s \in A_0$ and the second one, if it exists, belongs to I.

4.7.3   Family of quadratic polynomials

The behavior of the sequence $\{\tau_0, \tau_1, \tau_2, \ldots\}$, where $\tau_{n+1} = G(\tau_n)$, will be investigated for $\tau_0 \in I_0$. For numerous values of ε and k the map

$$f(\tau; \varepsilon, k) \stackrel{\text{def}}{=} G^k(\tau; \varepsilon) - 2\pi \equiv G^k(\tau - 2\pi; \varepsilon), \quad \tau \in I_0(\varepsilon) \tag{4.93}$$

is described by a quadratic polynomial.

Indeed, by (4.87) we have $f(\tau; \varepsilon, 1) = \varepsilon - b(\tau - c)^2$ for $\tau \in I_0(\varepsilon)$, and according to (4.88), there is

$$f(\tau; \varepsilon, k) = \begin{cases} q^{k-1}\big(\varepsilon - b(\tau - c)^2\big) & \text{if } \varepsilon q^{k-1} \le c + \sqrt{\varepsilon/b} \\ G\Big(q^{k-2}\big(\varepsilon - b(q\tau - c)^2\big)\Big) & \text{if } \varepsilon q^{k-2} \le c + \sqrt{\varepsilon/b} < \varepsilon q^{k-1} \end{cases}$$

$$\tag{4.94}$$

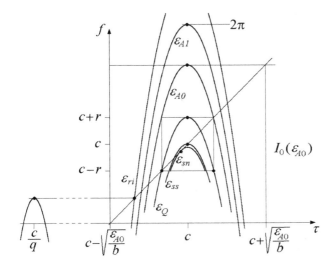

Fig. 4.38   Graphs of the function $f(\tau; \varepsilon, k)$, $\tau \in I_0(\varepsilon_{A0})$ for a fixed natural k and for $\varepsilon = \varepsilon_{sn}, \varepsilon_{ss}, \varepsilon_Q, \varepsilon_{A0}, \varepsilon_{A1}, \varepsilon_{ri}$.

for $\tau \in I_0(\varepsilon)$.

The graphs of the function $f(\tau; \varepsilon, k)$ are shown in Fig.4.38 for a fixed k, for small values of $|\tau - c|$ and for selected values of ε:

$$\varepsilon_{sn}(k) < \varepsilon_{ss}(k) < \varepsilon_Q(k) < \varepsilon_{A0}(k) < \varepsilon_{A1}(k) < \varepsilon_{ri}(k), \qquad (4.95)$$

for which qualitative properties of dynamics are changing.

The dynamics described by the one-parameter family of *quadratic polynomials*

$$f_a(x) = a - x^2 \quad \text{where} \quad a \in [-0.25, 2] \qquad (4.96)$$

is well known (see Refs. [15], [39]). The steady-state trajectories $\{f_a^n(0)\}$ of a critical point $x_0 = 0$ for $n = 400, \dots, 1000$ are shown in Fig.4.39 versus the parameter a. The same picture can be obtained for almost all initial values x_0 which satisfy the condition $|x_0| < \frac{1}{2}\left(1 + \sqrt{1 + 4a}\right)$. If x_0 is out of this interval then $\lim_{n \to \infty} x_n = -\infty$ for almost all values of x_0.

All qualitative properties of this dynamics (the saddle-node and period doubling bifurcations, the Feigenbaum cascade, the existence of invariant measures, the positive Liapunov exponent etc.) are the same as in Section 4.6. For $a = -0.25$, the fixed point $x = -0.5$ of the map (4.96) arises as a result of the saddle-node bifurcation. For $a = 0$, the fixed point $x = 0$

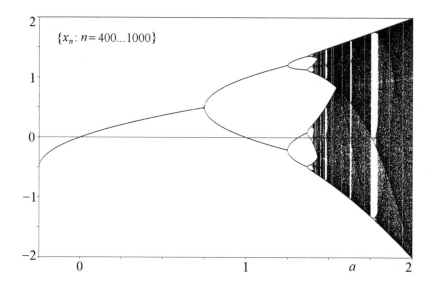

Fig. 4.39   The steady-state trajectories $x_{n+1} = a - x_n^2$ for $n = 400, ..., 1000$ versus a.

is superstable. For $a = 0.75$, there appears the first period doubling bifurcation which initiates the Feigenbaum cascade. For $a = 1.74996...$ the first periodic orbit of period 3 arises with the consequences given by the Li & Yorke theorem. For $a = 2$, the density of the invariant measure is $p_\infty(\xi) = \left(\pi\sqrt{4 - \xi^2}\right)^{-1}$, and the Liapunov exponent is $\lambda = \ln 2$. For $a > 2$ almost all trajectories tend to $-\infty$.

For a fixed k, if $\varepsilon \in (\varepsilon_{sn}, \varepsilon_Q)$ then dynamics described by the map (4.93) looks as dynamics of the family of quadratic polynomials shown in Fig.4.39 for $a \in (-0.25, 2)$. If $\varepsilon \in (\varepsilon_{A0}, \varepsilon_{A1})$ then trajectories of both critical points are attracted to the same fixed point τ_s. For ε_{ri} we have a bifurcation of rotation interval. If $\varepsilon \in (\varepsilon_Q, \varepsilon_{A0})$ or if $\varepsilon \in (\varepsilon_{A1}, \varepsilon_{ri})$, then the dynamics of the map (4.93) can be much more complicated. Next subsection describes these properties more precisely.

4.7.4   Dynamics restricted to the set I_0

Let k be a fixed natural number. Now we present how the values (4.95) depend on the critical point c and derivative q of the map G at its unstable fixed point.

1° For $\varepsilon = \varepsilon_{sn}$ a neutral fixed point τ^* of the map (4.94) arises as a result of the saddle-node bifurcation. The values ε_{sn} and τ^* satisfy the equations

$$f(\tau^*; \varepsilon_{sn}, k) = \tau^*, \qquad \frac{df}{d\tau}(\tau^*; \varepsilon_{sn}, k) = 1, \qquad \tau^* \in I_0,$$

and by (4.94) we get

$$\varepsilon_{sn} = \varepsilon_{ss}\left(1 - \frac{\varepsilon_{ss}}{4bc^2}\right), \qquad \tau^* = c\left(1 - \frac{\varepsilon_{ss}}{2bc^2}\right). \tag{4.97}$$

$2°$ For $\varepsilon = \varepsilon_{ss} = cq^{1-k}$ there exists the superstable fixed point $\tau = c$ of the map $f(\tau; \varepsilon, k)$. It was discussed in Section 4.7.2.

$3°$ For $\varepsilon = \varepsilon_Q$ there exist in I_0 two intervals $[c-r, c]$ and $[c, c+r]$ such that the quadratic polynomial $f(\tau; \varepsilon, k)$ maps each of them onto $[c-r, c+r]$. From the equations

$$f(c - r; \varepsilon_Q, k) = f(c + r; \varepsilon_Q, k) = c - r, \qquad f(c; \varepsilon_Q, k) = c + r,$$

we get

$$\varepsilon_Q = \varepsilon_{ss}\left(1 + \frac{2\varepsilon_{ss}}{bc^2}\right), \qquad r = \frac{2\varepsilon_{ss}}{bc}. \tag{4.98}$$

For $\varepsilon \in [\varepsilon_{sn}, \varepsilon_Q]$ the formula $\tau_{n+1} = f(\tau_n; \varepsilon, k)$, where $|\tau_n - c| < r$, can be reduced to the form

$$x_{n+1} = a(\varepsilon) - x_n^2, \qquad \text{where} \quad a(\varepsilon) = \frac{bc^2(\varepsilon - \varepsilon_{ss})}{\varepsilon_{ss}^2},$$

by replacing τ by $c + \varepsilon_{ss}x/bc$.

The parameter $a(\varepsilon)$ is an increasing function of ε, and $a(\varepsilon_{sn}) = -0.25$, $a(\varepsilon_{ss}) = 0$, $a(\varepsilon_Q) = 2$. So, for $\varepsilon \in (\varepsilon_{sn}, \varepsilon_Q)$ and for almost all initial values τ_0 satisfying the condition $|\tau_0 - c| < \frac{\varepsilon_{ss}}{2bc}\left(1 + \sqrt{1 + 4a(\varepsilon)}\right)$, the graph of the steady-state trajectories of the map $f(\tau; \varepsilon, k)$ looks like in Fig.4.39.

$4°$ For $\varepsilon \in (\varepsilon_{A0}, \varepsilon_{A1})$ the critical value $f(c; \varepsilon, k)$ belongs to the immediate attractive interval of the stable fixed point τ_s, i.e. $f(c; \varepsilon, k) \in (c + \sqrt{\varepsilon/b}, 2\pi)$. The trajectories of both critical points are attracted to the same fixed point, and if G has negative Schwarzian derivative then the point τ_s is the unique stable periodic trajectory. The values $\varepsilon_{A0}, \varepsilon_{A1}$ satisfy the equations

$$f(c; \varepsilon_{A0}, k) = c + \sqrt{\varepsilon_{A0}/b}, \qquad f(c; \varepsilon_{A1}, k - 1) = c - \sqrt{\varepsilon_{A1}/b}.$$

Hence

$$\varepsilon_{A0} = \varepsilon_{ss}\left(1 + \sqrt{\frac{\varepsilon_{ss}}{bc^2}\left(1 + \frac{\varepsilon_{ss}}{4bc^2}\right)} + \frac{\varepsilon_{ss}}{2bc^2}\right), \tag{4.99}$$

$$\varepsilon_{A1} = \varepsilon_{ss}q\left(1 - \sqrt{\frac{\varepsilon_{ss}q}{bc^2}\left(1 + \frac{\varepsilon_{ss}q}{4bc^2}\right)} + \frac{\varepsilon_{ss}q}{2bc^2}\right). \tag{4.100}$$

For $\varepsilon \in (\varepsilon_{A0}, \varepsilon_{A1})$ the trajectory $\{G^n(c-2\pi; \varepsilon)\}$ and trajectories starting from a sufficiently small neighborhood of the critical point $c - 2\pi$ tend to the fixed point $\tau_s = G(\tau_s)$.

5° The value ε_{ri} of the parameter ε is the bifurcation point of the rotation interval such that

$$\rho\big(f_+(\tau; \varepsilon, k)\big) = 0 \ \text{ for } \ \varepsilon \in (\varepsilon_{sn}, \varepsilon_{ri}) \ \text{ and } \ \rho\big(f_+(\tau; \varepsilon, k)\big) > 0 \ \text{ for } \ \varepsilon > \varepsilon_{ri}.$$

For $\varepsilon = \varepsilon_{ri}$ there exists in the interval $I_0(\varepsilon_{ri})$ such a fixed point $\tau^* < c$ of the map $f(\tau; \varepsilon, k)$, that

$$\tau^* = f(\tau^*; \varepsilon_{ri}, k) = f(cq^{-1}; \varepsilon_{ri}, k),$$

and for $\varepsilon > \varepsilon_{ri}$ the inequality $f_+(\tau; \varepsilon, k) > \tau$ holds for all τ (Fig.4.38). The solution of the above equations is

$$\varepsilon_{ri} = \varepsilon_{ss}q \left(1 - \sqrt{\frac{\varepsilon_{ss}(q-1)}{bc^2}\left(1 + \frac{\varepsilon_{ss}(q-1)}{4bc^2}\right)} + \frac{\varepsilon_{ss}(q-1)}{2bc^2} \right) \quad (4.101)$$

and $\tau^* = \varepsilon_{ri}q^{k-2}$.

6° If $\varepsilon \in \big(\varepsilon_Q(k), \varepsilon_{A0}(k)\big)$ or if $\varepsilon \in \big(\varepsilon_{A1}(k), \varepsilon_{sn}(k-1)\big)$, then the properties of the trajectory $\{G^n(c; \varepsilon)\}$ of the critical point c are more complicated. We are dealing with the first interval only because the properties given below are similar for both intervals.

If k is a fixed integer and $\varepsilon \in (\varepsilon_Q, \varepsilon_{A0})$, then $G^k(c - 2\pi; \varepsilon) = \varepsilon q^{k-1} \in$ $\in \big(c+r, \ c+\sqrt{\varepsilon_{A0}/b}\big)$. Let us denote

$$\varepsilon^1 = G^{k+1}(c - 4\pi; \varepsilon) \equiv G^{k+1}(c - 2\pi; \varepsilon) - 2\pi.$$

Evidently, ε^1 depends on ε and if ε increases from ε_Q to ε_{A0} then ε^1 decreases from $(\varepsilon - br^2)$ to zero. The value ε^1 belongs almost everywhere to one of the subsets A_n or I_n defined in Section 4.7.1.

If $\varepsilon^1 \in A = \cup A_n$, then $\lim_{n\to\infty} G^n(c - 4\pi; \varepsilon) = \tau_s$.

If $\varepsilon^1 \in I_n$ for a natural number $n > k$, then $G^{k+1+n}(c-4\pi; \varepsilon) = \varepsilon^1 q^n \in I_0$, and investigation of the properties of the trajectory is similar to that presented above in points 1°,...,5°. In particular, in the interval $\big(\varepsilon_Q(k), \varepsilon_{A0}(k)\big)$ there exists an infinite number of such values ε_n for which

$$G^{k+1+n}(c - 4\pi; \varepsilon_n) = c,$$

and G has superstable trajectory of the type $2/(k + 1 + n)$.

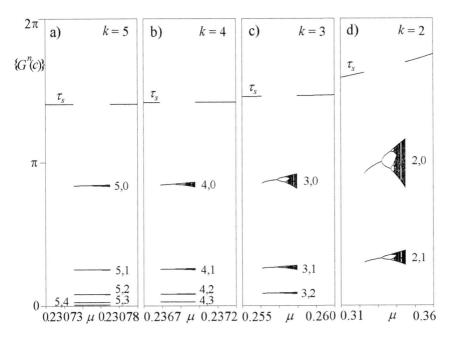

Fig. 4.40 Graphs of steady-state trajectories of the critical point c of the map (4.102) versus μ for $z = -\arcsin(2\pi\mu/a)$ and for $a = 2.45$.

4.7.5 *Asymptotic properties for* $\varepsilon \rightarrow 0$

Some results of a numerical experiment are shown in Fig.4.40.

The steady-state trajectories of the critical point c of the map

$$G(\tau) = \tau + 2\pi\mu + a\sin(\tau + z),\ mod\,2\pi \qquad (4.102)$$

are calculated for $a = 2.45$ and for such a value of z, that zero is an unstable fixed point of this map. The graphs $\{G^n(c) : n = 1000, ..., 2000\}$ versus μ are shown in Fig.4.40 for four intervals of μ of lengths equal to $0.5 \cdot 10^{-k}$. These intervals contain such values of μ that $\varepsilon(\mu) = G(c) - 2\pi \in$ $\in [\varepsilon_{sn}(k), \varepsilon_Q(k)]$ for $k = 5, 4, 3, 2$.

Components of the graphs are indexed by two integers k, l, where $1/k$ is the rotation number of the trajectories, and $l = 0, 1, 2, ..., k-1$ is the number of interval I_l which contains the terms of the trajectory $\{G^{kn+l}(c)\}$ for $n = 0, 1, 2, ...$ Each of the components mentioned above looks like in Fig.4.39 but with a different scale. If the parameter μ will be replaced by

$\varepsilon(\mu)$ then the component indexed by the numbers k, l has the following size:

$$\text{width} \ = \ \frac{9}{4b}q^{2-2k} \quad \text{and} \quad \text{height} \ = \ \frac{4}{b}q^{1-k-l}. \qquad (4.103)$$

Indeed, the width of a component is equal to $|\varepsilon_Q(k) - \varepsilon_{sn}(k)|$, and the height, for $l = 0$, is equal to $2r$, where r is given by (4.98). These sizes are approximate for the map (4.102) and exact for the map which satisfies the conditions $1°, 2°, 3°$ given in Section 4.7.1.

Let ε_0 be a small positive number. For a random value of ε from the interval $(0, \varepsilon_0)$ the probability that the trajectory $\{G^m(c; \varepsilon)\}$ of the critical point does not tend to τ_s is of the order $O(\varepsilon_0)$. If an initial point of the trajectory is not the critical point but a random value from the interval $(0, 2\pi)$, then this probability is of the order $O\left(\varepsilon_0^2\right)$. So, if the parameters of the DPLL system leave a frequency locking region D_N crossing over the *rib*-lines then synchronization preserves with significant probability. This is not true if the *snb*-lines are crossed (see Fig.4.35).

Chapter 5

Two-dimensional discrete-time Phase-Locked Loop

5.1 Description of the DPLL system by a two-dimensional map

The system discussed in this chapter differs from the one-dimensional system presented in Section 4.1 in two following points:

1) At the moment of sampling t_n the controlling voltage $u(t)$ satisfies the condition

$$u(t_n + 0) = \lambda u(t_n - 0) + (1 - \lambda)U_{\text{inp}}(\omega t_n), \tag{5.1}$$

where $\lambda \in (0, 1)$ or, equivalently, the *jump* of $u(t)$ is defined by

$$u(t_n + 0) - u(t_n - 0) = (1 - \lambda)\big(U_{\text{inp}}(\omega t_n) - u(t_n - 0)\big).$$

2) The controlling voltage $u(t)$ is not a constant value for $t \in (t_n, t_{n+1})$, but it exponentially decreases:

$$u(t) = u(t_n + 0)e^{-b\Omega(t-t_n)} \quad \text{for} \quad t \in (t_n, t_{n+1}), \tag{5.2}$$

where $b > 0$.

This model seems to be more realistic. The S&H unit shown in Fig.4.1 is now a *sample-and-filter* unit. For $\lambda = 0$ and $b = 0$ it reduces to the unit discussed in Chapter 4

Using the equation of the voltage controlled oscillator

$$\frac{d\theta}{dt} = \Omega\left(1 + \frac{u(t)}{V}\right)$$

and (5.2) we obtain

$$2\pi = \theta(t_{n+1}) - \theta(t_n) = \Omega(t_{n+1} - t_n) + u(t_n + 0)\frac{1}{bV}\left(1 - e^{-b\Omega(t_{n+1}-t_n)}\right),$$

$$u(t_{n+1} - 0) = u(t_n + 0)e^{-b\Omega(t_{n+1}-t_n)}.$$

Let us introduce dimensionless quantities

$$\tau_n = \omega t_n, \quad x_n = \frac{u(t_n - 0)}{V}, \quad z_n = \frac{u(t_n + 0)}{V}, \quad \gamma_n = \frac{\Omega}{\omega}(\tau_{n+1} - \tau_n).$$

Then, the equations of the system take the form

$$z_n = \lambda x_n + (1 - \lambda)\frac{1}{V}U_{\text{inp}}(\tau_n), \tag{5.3}$$

$$2\pi = \gamma_n + z_n\frac{1}{b}\left(1 - e^{-b\gamma_n}\right), \tag{5.4}$$

$$x_{n+1} = z_n e^{-b\gamma_n}. \tag{5.5}$$

Properties of the system can be described by the sequence

$$(\tau_0, x_0), \quad (\tau_1, x_1), \quad (\tau_2, x_2), \quad (\tau_3, x_3), \ldots \tag{5.6}$$

of time τ_n and controlling voltage x_n (both dimensionless) at the moments of sampling for which $\theta = 2\pi n$. According to (5.3)–(5.5), the sequence is defined by the recurrent formula $(\tau_{n+1}, x_{n+1}) = F(\tau_n, x_n)$ given by

$$\tau_{n+1} = \tau_n + \frac{\omega}{\Omega} \cdot g_b(z_n),$$

$$x_{n+1} = h_b(z_n), \quad \text{where} \quad z_n = \lambda x_n + (1 - \lambda)\frac{1}{V}U_{\text{inp}}(\tau_n).$$

The functions g_b and h_b are defined as follows. The formula $\gamma_n = g_b(z_n)$ gives a solution of equation (5.4) or, equivalently, $\gamma = g_b(z)$ is the inverse function of

$$z = g_b^{-1}(\gamma) \stackrel{\text{def}}{=} b\frac{2\pi - \gamma}{1 - e^{-b\gamma}}, \quad \text{for} \quad b > 0. \tag{5.7}$$

The function h_b is defined by

$$h_b(z) = ze^{-bg_b(z)} \equiv z + b(g_b(z) - 2\pi). \tag{5.8}$$

The graphs of both functions are shown in Fig.5.1 for several values of b.

If $b > 0$ then both functions are analytic for all real z. Their derivatives are

$$\frac{dg_b}{dz} = -\frac{1}{b} \cdot \frac{1 - e^{-b\gamma}}{1 + ze^{-b\gamma}}, \quad \frac{dh_b}{dz} = \frac{1 + z}{z + e^{b\gamma}}, \quad \text{where} \quad \gamma = g_b(z).$$

The function g_b decreases and takes positive values. For $z > -1$ the function h_b satisfies the conditions: $h_b(z) > -1$ and $0 < \frac{h_b(z)}{z} < 1$, and moreover $0 < \frac{dh_b(z)}{dz} < 1$. For $z = 0$ we have

$$g_b(0) = 2\pi, \quad g_b'(0) = -\frac{1}{b}\left(1 - e^{-2\pi b}\right), \quad h_b(0) = 0, \quad h_b'(0) = e^{-2\pi b}.$$

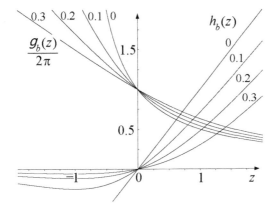

Fig. 5.1   Graphs of the functions g_b and h_b for $b = 0, 0.1, 0.2, 0.3$.

For $b = 0$ we define

$$g_0(z) = \frac{2\pi}{1+z} \quad \text{and} \quad h_0(z) = z \quad \text{for} \quad z > -1. \qquad (5.9)$$

For the standard input signal $U_{\text{inp}}(\tau) = AV \sin \tau$, where $0 < A < 1$, the map $F : (\tau, x) \to (\tau', x')$ takes the form

$$\tau' = \tau + \mu\, g_b(z), \qquad (5.10)$$

$$x' = h_b(z), \quad \text{where} \quad z = \lambda x + (1 - \lambda) A \sin \tau. \qquad (5.11)$$

It depends on four parameters: b, λ, A and $\mu = \omega/\Omega$.

Let the domain of the map F be $D_A = \{\tau, x : |x| < A\}$, or the corresponding subset of cylinder $S \times R$, where $\tau \bmod 2\pi$ is a cyclic variable. It is not hard to see that $F(D_A) \subset D_A$, and F has the inverse map F^{-1}:

$$\tau = \tau' - \mu\, g_b\left(h_b^{-1}(x')\right),$$

$$x = \frac{1}{\lambda} h_b^{-1}(x') - \frac{1-\lambda}{\lambda} A \sin\left(\tau' - \mu\, g_b\left(h_b^{-1}(x')\right)\right),$$

contrary to the case $\lambda = 0$, where T^{-1} does not exist.

Let us consider the map $F = F_2 \circ F_1$ as a composition of two maps, each of them changes only one coordinate:

$$F_1 : (\tau, x) \to (\tau, x') \quad \text{and} \quad F_2 : (\tau, x') \to (\tau', x'). \qquad (5.12)$$

The map F_1 (for $\tau = const.$) is a contracting one in the set $|x| < A$. Its Lipschitz constant is smaller then $\lambda h_b'(A) \in (0, 1)$. For any fixed τ, the map

$$x' = h_b\big(\lambda x + A(1 - \lambda) \sin \tau\big) \qquad (5.13)$$

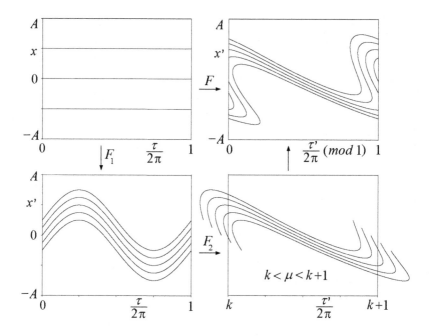

Fig. 5.2   Geometrical properties of the maps F_1 and F_2.

has exactly one fixed point $x_c = x_c(\tau)$. The map F_1 contracts the set D_A to a neighborhood of its invariant curve $x_c = x_c(\tau)$ (see Fig.5.2).

The map F_2 (for $x' = const.$) shifts a point (τ, x') along τ-axis by the distance

$$\tau' - \tau = \mu \, g_b \left(h_b^{-1}(x') \right), \tag{5.14}$$

which is decreasing function of the parameter x'. In particular, the right-hand side of (5.14) takes the value $2\pi\mu$ for $x' = 0$.

The set $F(D_A)$ can be folded (Fig.5.2), and then the dynamics defined by F complicates (Smale's horseshoes can exist). The equality $F(\tau + 2\pi, x) = F(\tau, x) + 2\pi$ implies that the set $F(D_A)$ can be considered as a subset of cylinder $S \times R$.

Let mes$D$ be the area of a domain $D \subset D_A$. The area of $F(D)$ is

$$\text{mes}\, F(D) = \iint_{F(D)} d\tau' dx' = \iint_D |J(\tau, x)| d\tau dx \le \max_D |J(\tau, x)| \, \text{mes}\, D,$$

where $J(\tau, x)$ is the Jacobian of the map (5.10)–(5.11), i.e. the determinant of the Jacobian matrix (derivative of F):

$$F'(\tau, x) = \begin{bmatrix} 1 + \mu\, g_b'(z)\frac{\partial z}{\partial \tau} & \mu\, g_b'(z)\frac{\partial z}{\partial x} \\ h_b'(z)\frac{\partial z}{\partial \tau} & h_b'(z)\frac{\partial z}{\partial x} \end{bmatrix}, \tag{5.15}$$

where $z = \lambda x + (1 - \lambda)A \sin \tau$.

We have $J(\tau, x) = \lambda h_b'(z)$, and if $|x| < A$ then $0 < J(\tau, x) \le \lambda h_b'(A) \le$
$\le \lambda < 1$. In particular, $J(\tau, x) = \lambda$ for $b = 0$. For each domain $D \subset D_A$,
the area $\mathrm{mes}F(D)$ is smaller then $\mathrm{mes}D$. Every invariant set $\Lambda = F(\Lambda)$
of the map F has measure (area) equal to zero. An example of the sets
$F(D_A)$ and $F^3(D_A)$ is shown in Fig.5.3 for fixed values of the parameters:
b, λ, A, μ.

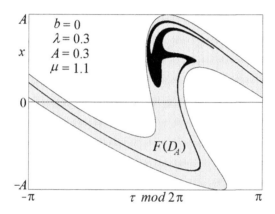

Fig. 5.3   An example of the sets $F(D_A)$ (grey domain) and $F^3(D_A)$ (black domain).

Our purpose is to investigate trajectories $\{F^n(\tau_0, x_0)\}_{n=0}^{\infty}$ and attractive
invariant sets Λ of the map F. The following different sets Λ can exist:
stable fixed points, stable periodic orbits, stable invariant curves or strange
attractors. They will be discussed in succeeding sections of this chapter.

5.2   Stable periodic orbits

We call (τ_*, x_*) the periodic point of the type n/m of the map F, if

$$F^m(\tau_*, x_*) = (\tau_* + 2\pi n, x_*) \tag{5.16}$$

and if m is the smallest natural number for which (5.16) holds for an integer n. For example, if (τ_*, x_*) satisfies (5.16) for $(n, m) = (12, 6)$ and does
not it for $(6, 3)$ and for $(2, 1)$ then (τ_*, x_*) is of the type $12/6$.

The sequence of the points

$$(\tau_*, x_*),\ F(\tau_*, x_*),\ F^2(\tau_*, x_*),\ \ldots F^{m-1}(\tau_*, x_*)$$

is called the periodic orbit of the point (τ_*, x_*), where τ_* is understood as a value of a cyclic $(mod\, 2\pi)$ variable.

If (τ_*, x_*) is a periodic point of the type n/m, then for initial values $t_0 = \tau_*/\omega$, $u(t_0 - 0) = V x_*$, the controlling voltage $u(t)$ and the output signal $U_{\text{out}}(\theta(t))$ (with $\theta(t_0) = 0$) of DPLL system are periodic of the period $2\pi n/\omega$, and the phase $\theta(t)$ increases by $2\pi m$ during this period. Examination of periodic output signals is reduced to the solving of equations (5.16) for different natural numbers n and m. If $m \geq 2$, then only numerical methods can be used, but for $m = 1$ elementary methods are sufficient.

5.2.1   *Periodic points of the type $n/1$*

If (τ_*, x_*) is a periodic point of the type $n/1$, then the following equations:

$$g_b(z_*) = \frac{2\pi n}{\mu}, \qquad x_* = h_b(z_*), \qquad z_* = \lambda x_* + (1 - \lambda) A \sin \tau_* \qquad (5.17)$$

are satisfied.

Hence, for $b > 0$ we have

$$z_* = g_b^{-1}(2\pi n/\mu) = \frac{\mu - n}{\mu} \cdot \frac{2\pi b}{1 - e^{-2\pi bn/\mu}}, \qquad (5.18)$$

$$x_* = h_b(z_*) = \frac{\mu - n}{\mu} \cdot \frac{2\pi b\, e^{-2\pi bn/\mu}}{1 - e^{-2\pi bn/\mu}}, \qquad (5.19)$$

$$A \sin \tau_* = H_*, \qquad (5.20)$$

where

$$H_* = \frac{z_* - \lambda x_*}{1 - \lambda} = \frac{\mu - n}{\mu} \cdot \frac{2\pi b\left(1 - \lambda e^{-2\pi bn/\mu}\right)}{(1 - \lambda)(1 - e^{-2\pi bn/\mu})}. \qquad (5.21)$$

Let us mention that x_* does not depend on λ.

Solutions of the equations (5.17) exist if and only if $A \geq |H_*|$. If $A > |H_*|$ then exactly two solutions (τ_*, x_*) and $(\pi - \tau_*, x_*)$ exist on cylinder $S \times R$. For fixed values A and μ, the condition $A > |H_*|$ can be satisfied for several numbers n. For example, if $b = 0$ then

$$z_* = x_* = A \sin \tau_* = H_* \equiv \frac{\mu - n}{n} \qquad (5.22)$$

and there exist periodic points of all types $n/1$, for which

$$\frac{\mu}{1 + A} < n < \frac{\mu}{1 - A}. \qquad (5.23)$$

An example of the regions of existence of periodic points of the type $n/1$ is shown in Fig.5.4 for several integers n.

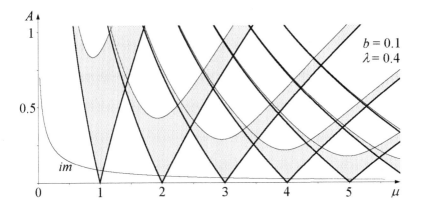

Fig. 5.4   Regions of existence of periodic points and attracting periodic points (grey domains) of the type $n/1$. Below the line im there exists a stable invariant manifold.

5.2.2   Stability of fixed points

Let a map $G(\tau, x) = \big(G_1(\tau, x), G_2(\tau, x)\big)$ have continuous partial derivatives of the second order in a neighborhood of a fixed point (τ_*, x_*) of the map $G(\tau, x)$. By Taylor's formula we have

$$
\begin{bmatrix} G_1(\tau, x) - \tau_* \\ G_2(\tau, x) - x_* \end{bmatrix} = \begin{bmatrix} \frac{\partial G_1}{\partial \tau}(\tau_*, x_*) & \frac{\partial G_1}{\partial x}(\tau_*, x_*) \\ \frac{\partial G_2}{\partial \tau}(\tau_*, x_*) & \frac{\partial G_2}{\partial x}(\tau_*, x_*) \end{bmatrix} \cdot \begin{bmatrix} \tau - \tau_* \\ x - x_* \end{bmatrix} + \begin{bmatrix} O(r^2) \\ O(r^2) \end{bmatrix}, \quad (5.24)
$$

where $r^2 = (\tau - \tau_*)^2 + (x - x_*)^2$.

The Jacobian matrix will be denoted by

$$
G'(\tau_*, x_*) = \begin{bmatrix} a & b \\ c & d \end{bmatrix} = M \begin{bmatrix} \nu_+ & 0 \\ 0 & \nu_- \end{bmatrix} M^{-1}, \quad (5.25)
$$

where ν_+, ν_- are the eigenvalues of the matrix $G'(\tau_*, x_*)$

$$
\nu_\pm = \frac{1}{2}\left(a + d \pm \sqrt{(a+d)^2 - 4(ad - bc)}\right). \quad (5.26)
$$

The columns of the matrix M are the eigenvectors of $G'(\tau_*, x_*)$. For example,

$$
M = \begin{bmatrix} b, & b \\ \nu_+ - a, & \nu_- - a \end{bmatrix} \quad \text{or} \quad M = \begin{bmatrix} \nu_+ - d, & \nu_- - d, \\ c, & c \end{bmatrix} \quad (5.27)
$$

respectively for $b \neq 0$ or $c \neq 0$, (if $b = c = 0$ then M is a unit matrix). The formulae (5.25) and (5.27) are correct for $\nu_- \neq \nu_+$ only (the case $\nu_- = \nu_+$ is not discussed here).

Let us consider the sequence of points

$$(\tau_0, x_0),\ (\tau_1, x_1),\ (\tau_2, x_2),\ldots \quad \text{where}\ (\tau_k, x_k) = G^k(\tau_0, x_0). \tag{5.28}$$

As long as the points of the sequence remain in a small neighborhood of the fixed point (τ_*, x_*), the map G can be linearized and the approximate equality

$$\begin{bmatrix} \tau_k - \tau_* \\ x_k - x_* \end{bmatrix} \approx M \begin{bmatrix} \nu_+^k & 0 \\ 0 & \nu_-^k \end{bmatrix} M^{-1} \begin{bmatrix} \tau_0 - \tau_* \\ x_0 - x_* \end{bmatrix} \tag{5.29}$$

holds. Formula (5.29) suggests the following classification of fixed points (see Ref. [50]).

Definition 5.1. The fixed point (τ_*, x_*) of the map G is called:
- attracting   if $|\nu_+| < 1$ and $|\nu_-| < 1$,
- repelling   if $|\nu_+| > 1$ and $|\nu_-| > 1$,
- saddle-type if $|\nu_-| < 1 < |\nu_+|$ or $|\nu_+| < 1 < |\nu_-|$,
- neutral   if $\nu_- = -1$ or $\nu_+ = 1$ or $|\nu_-| = |\nu_+| = 1$.

If (τ_*, x_*) is an attracting fixed point, then the set

$$A(\tau_*, x_*) = \left\{ (\tau, x) : \lim_{k \to \infty} G^k(\tau, x) = (\tau_*, x_*) \right\}$$

is called the attractive domain of the fixed point.

It is evident, that

$$\nu_- \cdot \nu_+ = \det G'(\tau_*, x_*), \qquad \nu_- + \nu_+ = \operatorname{Tr} G'(\tau_*, x_*). \tag{5.30}$$

Figure 5.5 shows how the eigenvalues ν_-, ν_+ depend on the determinant and on the trace of the matrix $G'(\tau_*, x_*)$.

The fixed point (τ_*, x_*) is attracting if and only if

$$|\operatorname{Tr} G'(\tau_*, x_*)| - 1 < \det G'(\tau_*, x_*) < 1. \tag{5.31}$$

Above the parabola marked by dashed line in Fig.5.5 eigenvalues are complex conjugate and fixed point is focus. Below this parabola eigenvalues are real numbers and fixed point is node or saddle-type.

Neutral fixed points occur on the lines (solid lines in Fig.5.5) which separate the regions of attracting, repelling and saddle-type fixed points.

Now let us introduce the notion of locally invariant curves in a neighborhood of the node and the saddle-type fixed point (τ_*, x_*). Eigenvalues and eigenvectors of the matrix $G'(\tau_*, x_*)$ are real. Let us introduce the system of coordinates u, v:

$$\begin{bmatrix} u \\ v \end{bmatrix} = M^{-1} \begin{bmatrix} \tau - \tau_* \\ x - x_* \end{bmatrix}, \quad \text{or inversely} \quad \begin{bmatrix} \tau \\ x \end{bmatrix} = \begin{bmatrix} \tau_* \\ x_* \end{bmatrix} + M \begin{bmatrix} u \\ v \end{bmatrix}. \tag{5.32}$$

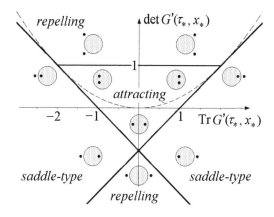

Fig. 5.5   Regions of existence of attracting, repelling and saddle-type fixed points. Locations of the eigenvalues with respect to the unit circle are also shown.

According to (5.29), for the linearized map we have

$$\begin{bmatrix} u_k \\ v_k \end{bmatrix} = \begin{bmatrix} \nu_+^k & 0 \\ 0 & \nu_-^k \end{bmatrix} \begin{bmatrix} u_0 \\ v_0 \end{bmatrix}.$$

Hence $v_k = C(u_k)^\alpha$ for every natural k, where $\alpha = \frac{\ln|\nu_-|}{\ln|\nu_+|}$ and where C depends on the initial point (u_0, v_0), with $u_0 \neq 0$. If $u_0 = 0$, then $u_k = 0$. In the system of coordinates u, v the family of lines:

$$u = 0 \quad \text{and} \quad v = Cu^\alpha, \quad \text{where} \quad \alpha = \frac{\ln|\nu_-|}{\ln|\nu_+|}, \qquad (5.33)$$

with a parameter $C \in (-\infty, +\infty)$, is called the family of *locally invariant curves*.

The pictures of these curves in the system of coordinates τ, x are shown in Fig.5.7 for a node (where $\alpha > 0$) and for a saddle-type point (where $\alpha < 0$). Dotted lines are described by the equation $u = 0$.

The saddle-type fixed point is the intersection of two curves L_+, L_- invariant with respect to G and tangent to the eigenvectors of the matrix $G'(\tau_*, x_*)$ at the fixed point (τ_*, x_*). Let ν_s and ν_u denote eigenvalues with absolute values smaller and greater then one. If $(\tau, x) \in L_+$ then $G^k(\tau, x) \to (\tau_*, x_*)$ for $k \to \infty$, as fast as a geometric progression with the ratio ν_s. If $(\tau, x) \in L_-$ then $G^k(\tau, x) \to (\tau_*, x_*)$ for $k \to -\infty$, as fast as a geometric progression with the ratio ν_u^{-1}.

5.2.3   Hold-in regions

The Jacobian of the map F defined by (5.10)–(5.11) satisfies the condition $0 < J(\tau, x) \leq \lambda$. Each point $(\tau_k, x_k) = F^k(\tau_*, x_*)$ of a periodic orbit of the type n/m is a fixed point of the map

$$G(\tau, x) = F^m(\tau, x) - (2\pi n, 0), \qquad (5.34)$$

and the Jacobian $\det G'(\tau_*, x_*)$ takes values from the interval $(0, \lambda^m]$. So, only the attracting and the saddle-type periodic points can exist (repelling points are not possible).

According to (5.31) and (5.15), the periodic point (τ_*, x_*) of the type $n/1$ of the map F is attracting if and only if $\cos \tau_* > 0$ and

$$|H_*| < A < \sqrt{H_*^2 + \left(\frac{2}{\mu} \cdot \frac{1 + \lambda h_b'(z_*)}{(1-\lambda) g_b'(z_*)} \right)^2}, \qquad (5.35)$$

where H_* is the value defined by (5.21). The domains where (5.35) is satisfied, so called hold-in regions, are marked in grey in Fig.5.4 for $n = = 1, ..., 5$.

If $A > |H_*|$ then exactly two periodic points (τ_*, x_*) and $(\pi - \tau_*, x_*)$ of the type $n/1$ exist on $S \times R$. One of them (this one for which $\cos \tau_* < 0$) is of the saddle-type. The second one is either attracting, if (5.35) holds, or saddle-type when (5.35) does not hold.

Let A be changing. If $A \to |H_*|$ then both periodic points approach $(\frac{1}{2}\pi, x_*)$ or $(-\frac{1}{2}\pi, x_*)$ and disappear (saddle-node bifurcation). If A exceeds the value of the right-hand side of (5.35), then the attracting point changes into saddle-type point, and a stable orbit of the type $2n/2$ branches off the fixed point (period doubling bifurcation). Both bifurcations are analogical to those described in Section 4.6 for one-dimensional maps.

In the case $b = 0$ the inequalities (5.35) take the form

$$\left| \frac{\mu - n}{n} \right| < A < \sqrt{\left(\frac{\mu - n}{n} \right)^2 + \left(\frac{\mu}{\pi n^2} \cdot \frac{1 + \lambda}{1 - \lambda} \right)^2}. \qquad (5.36)$$

For fixed values of parameters A, μ, λ, the attracting periodic points of the type $n/1$ can exist for several integers n, namely, for each n which satisfies inequality

$$\frac{1 + \lambda}{1 - \lambda} > n^2 \frac{\pi}{\mu} \sqrt{A^2 - \left(\frac{\mu - n}{n} \right)^2}. \qquad (5.37)$$

For example, if $A = 0.3$, $\mu = 6$, $\lambda = 0.7$, then (5.37) holds for $n = 5, 6, 8$. The stable output signal $U_{\text{out}}(\theta(t))$ of DPLL system can be periodic of the types $5/1$, or $6/1$, or $8/1$, but not $7/1$.

5.2.4   *Small values of* λ

For small values of λ the map (5.10)–(5.11) is strongly contracting in the direction of x-axis, and it maps the set $\{(\tau, x) : |x| < c\}$ where $0 < c < \infty$, onto a small neighborhood of the curve

$$\tau' = \alpha + \mu g_b(A \sin \alpha), \quad x' = h_b(A \sin \alpha), \quad \alpha \in (-\infty, +\infty). \tag{5.38}$$

If $b=0$, then we put $\{(\tau, x) : -1 + \frac{1}{c} < x < c\}$ in place of $\{(\tau, x) : |x| < c\}$.

We will prove that periodic points of the map F defined by (5.10)–(5.11) are well approximated by the periodic points of the one-dimensional map

$$T(\tau) = \tau + \mu \, g_b(A \sin \tau). \tag{5.39}$$

Theorem 5.1. *Let τ_0^* be a periodic point of the type n/m of the map* (5.39) *with the multiplier*

$$d = T'(\tau_0^*) \cdot T'\big(T(\tau_0^*)\big) \cdot \ldots \cdot T'\big(T^{m-1}(\tau_0^*)\big) \neq 1. \tag{5.40}$$

Let us denote $x_0^* = h_b\big(A \sin T^{m-1}(\tau_0^*)\big)$. Then:

(a) *There exists such λ_* that for all $\lambda \in (0, \lambda_*)$ in a neighborhood of (τ_0^*, x_0^*) the map F has exactly one periodic point $(\tau_\lambda^*, x_\lambda^*)$ of the type n/m and $(\tau_\lambda^*, x_\lambda^*) \to (\tau_0^*, x_0^*)$ as $\lambda \to 0$.*

(b) *If τ_0^* is a stable (or unstable) periodic point of T, then $(\tau_\lambda^*, x_\lambda^*)$ is an attracting (or saddle-type, respectively) point of F.*

(c) *If $A(\tau_0^*)$ is an attractive set of a stable point τ_0^*, then every compact set $D \subset A(\tau_0^*) \times R$ belongs to the attractive domain of the point $(\tau_\lambda^*, x_\lambda^*)$ for sufficiently small λ. If $b=0$, then we put $-1 < x < \infty$ in place of R.*

Proof. (a) The map (5.10)–(5.11) is of the form

$$F(\tau, x; \lambda) = \begin{bmatrix} T(\tau) + \lambda f_1(\tau, x; \lambda) \\ h_b(A \sin \tau) + \lambda f_2(\tau, x; \lambda) \end{bmatrix}, \tag{5.41}$$

where f_1, f_2 are smooth and bounded functions. Hence, for fixed numbers n, m, the function $H(\tau, x; \lambda) = F^m(\tau, x; \lambda) - (\tau + 2\pi n, x)$ takes the form

$$H(\tau, x; \lambda) = \begin{bmatrix} T^m(\tau) - (\tau + 2\pi n) + \lambda f_3(\tau, x; \lambda) \\ h_b\big(A \sin T^{m-1}(\tau)\big) - x + \lambda f_4(\tau, x; \lambda) \end{bmatrix}, \tag{5.42}$$

where f_3, f_4 are also smooth and bounded functions.

If τ_0^* is a periodic point of the type n/m of the map (5.39), then (τ_0^*, x_0^*) satisfies the equation $H(\tau, x; 0) = 0$, and the matrix

$$H'(\tau_0^*, x_0^*; 0) = \begin{bmatrix} d - 1, & 0 \\ \frac{d}{d\tau} h_b\big(A \sin T^{m-1}(\tau)\big), & -1 \end{bmatrix}_{\tau = \tau_0^*} \tag{5.43}$$

is nonsingular.

By the implicit function theorem, for sufficiently small λ, in a neighborhood of (τ_0^*, x_0^*) there exists exactly one solution $(\tau_\lambda^*, x_\lambda^*)$ of the equation $H(\tau, x; \lambda) = 0$. This solution is a continuous function of λ in a neighborhood of the point $\lambda = 0$ and it is a periodic point of the type n/m of the map F.

(b) The point $(\tau_\lambda^*, x_\lambda^*)$ is a fixed point of the map $G(\tau, x; \lambda) = H(\tau, x; \lambda) + (\tau, x)$. The eigenvalues $\nu_1(\lambda)$, $\nu_2(\lambda)$ of the matrix $G'(\tau, x; \lambda) = H'(\tau, x; \lambda) + I$ are continuous functions of λ in a neighborhood of the point $(\tau_0^*, x_0^*; 0)$. Moreover, $\nu_1(\lambda) \to d$ and $\nu_2(\lambda) \to 0$ as $\lambda \to 0$. If λ is sufficiently small, then the point $(\tau_\lambda^*, x_\lambda^*)$ is attracting for $|d| < 1$, and it is of saddle-type for $|d| > 1$.

(c) Let $K(r)$ denotes a circle of radius r with the center at the point (τ_0^*, x_0^*). Let $K(2\varepsilon)$ belong to the attractive sets of fixed points $(\tau_\lambda^*, x_\lambda^*)$ of the maps $G(\tau, x; \lambda)$ for all $\lambda \in [0, \lambda_1(\varepsilon)]$, where $\lambda_1(\varepsilon)$ is a positive number.

Let D be a compact set and let $D \subset A(\tau_0^*) \times R$. Then, there exists an integer k such that $G^k(\tau, x; 0) \in K(\varepsilon)$ for all $(\tau, x) \in D$. By continuity of G^k, there exists a positive $\lambda_2 < \lambda_1(\varepsilon)$ such that $G^k(\tau, x; \lambda) \in K(2\varepsilon)$ for all $\lambda \in [0, \lambda_2]$. Hence $G^n(\tau, x; \lambda) \to (\tau_0^*, x_0^*)$ as $n \to \infty$ for all $(\tau, x) \in D$ and $\lambda \in [0, \lambda_2]$.

This completes the proof. $\square$

5.3   Reduction to a one-dimensional system

In this section the conditions will be given, under which dynamics of two-dimensional DPLL is the same as dynamics of one-dimensional system.

Let us consider the map F defined by (5.10)–(5.11), i.e.

$$\tau' = \tau + \mu\, g_b\big(\lambda x + (1 - \lambda)A\sin\tau\big),$$
$$x' = h_b\big(\lambda x + (1 - \lambda)A\sin\tau\big).$$

Definition 5.2. A smooth curve

$$M = \{(\tau, x) : x = f(\tau), \tau \in R\}, \tag{5.44}$$

where f is a continuous function, is called the invariant manifold of the map F if $F(M) = M$ or, equivalently, if

$$F\big(\tau, f(\tau)\big) = \big(\tau', f(\tau')\big). \tag{5.45}$$

The invariant manifold is called stable if, for every point (τ, x) from a neighborhood of M, the points $(\tau_k, x_k) = F^k(\tau, x)$ belong to this neighborhood for $k \geq 1$ and tend to the curve M as $k \to \infty$.

The set

$$A(M) = \left\{ (\tau, x) : \lim_{k \to \infty} |x_k - f(\tau_k)| = 0 \right\} \qquad (5.46)$$

is called the attractive set of M. If $A(M)$ coincides with the whole domain of F, then the manifold M is called globally stable.

If the map F has the globally stable invariant manifold (5.44), then the *steady-state* dynamics of the DPLL system is described by the one-dimensional map

$$\tau' = T(\tau) \overset{\text{def}}{=} \tau + \mu\, g_b\big(\lambda f(\tau) + (1 - \lambda)A \sin\tau\big). \qquad (5.47)$$

It means that for an arbitrary (τ, x) and for $(\tau_k, x_k) = F^k(\tau, x)$ we have approximate equalities $\tau_k \approx T(\tau_{k-1})$ and $x_k \approx f(\tau_k)$ for sufficiently large integers k. If $(\tau, x) \in M$ then these equalities are not approximate but exact ones for all natural numbers k.

We prove that such invariant manifold exists for some regions of parameters μ and A.

5.3.1   Existence of an invariant manifold

Let l_g and l_h denote the Lipschitz constants of functions g_b and h_b in the interval $[-A, A]$, i.e.

$$l_g = \max_{|z| \le A} |g_b'(z)|, \qquad l_h = \max_{|z| \le A} |h_b'(z)|.$$

Theorem 5.2. *Let $\lambda \in (0, 1)$ and $l_h \le 1$. If*

$$A(1 - \lambda)\mu l_g \le \left(1 - \sqrt{\lambda l_h}\right)^2, \qquad (5.48)$$

then the map (5.10)–(5.11) has exactly one globally stable, invariant manifold (5.44). The function f is bounded: $\sup_\tau |f(\tau)| \le A$, *periodic:* $f(\tau + 2\pi) = f(\tau)$ *and it has the bounded Lipschitz constant:*

$$\sup_{\tau_1, \tau_2} \left| \frac{f(\tau_1) - f(\tau_2)}{\tau_1 - \tau_2} \right| \le r_*, \quad \text{where } r_* = \frac{1 - \mu A(1 - \lambda)l_g - \lambda l_h}{2\mu\lambda l_g}. \qquad (5.49)$$

The one-dimensional map T defined by (5.47) is an increasing function.

Proof.   We start with the observation that for every point (τ, x) there exists an integer k such that $F^k(\tau, x)$ belongs to $D_A = \{(\tau, x) : |x| \le A\}$. Let $B(r)$ denote the set of continuous periodic functions $f(\tau + 2\pi) = f(\tau)$ which satisfy the conditions

$$\sup_\tau |f(\tau)| \le A, \qquad \sup_{\tau_1, \tau_2} \left| \frac{f(\tau_1) - f(\tau_2)}{\tau_1 - \tau_2} \right| \le r. \qquad (5.50)$$

Let the map F transform the curve $x = f(\tau)$ onto the curve $x' = f_F(\tau')$. This transformation we denote by $f_F = F(f)$.

We precede the proof of Theorem 5.2 by two lemmas.

Lemma 5.1. *If the condition* (5.48) *is satisfied then the map F transforms the set $B(r_*)$ into itself, where r_* is defined by* (5.49).

Proof. [Proof of Lemma] If $f \in B(r)$ then $\sup\limits_{\tau'} |f_F(\tau')| \leq A$ because $l_h \leq 1$. An easy computation shows that

$$\frac{df_F(\tau')}{d\tau'} = \frac{\frac{dx'(\tau)}{d\tau}}{\frac{d\tau'(\tau)}{d\tau}} = \frac{h_b'\big(z(\tau)\big)\frac{dz(\tau)}{d\tau}}{1 + \mu g_b'\big(z(\tau)\big)\frac{dz(\tau)}{d\tau}}, \tag{5.51}$$

where $z(\tau) = \lambda f(\tau) + (1 - \lambda)A\sin\tau$. If the derivatives of the functions $f(\tau)$ and $f_F(\tau')$ are replaced by increment ratios, then we obtain

$$\sup\limits_{\tau_1, \tau_2} \left| \frac{f_F(\tau_1) - f_F(\tau_2)}{\tau_1 - \tau_2} \right| \leq R(r), \tag{5.52}$$

where

$$R(r) = \frac{l_h\big(\lambda r + (1 - \lambda)A\big)}{1 - \mu l_g\big(\lambda r + (1 - \lambda)A\big)}. \tag{5.53}$$

It is easily seen that the condition (5.48) is necessary and sufficient for the existence of a positive solution of the inequality $R(r) \leq r$. The number r_* defined by (5.49) is a solution of this inequality. $\qquad\square$

Lemma 5.2. *If the condition* (5.48) *is satisfied then the map F defined on the set $B(r_*)$ satisfies the Lipschitz condition*

$$\sup\limits_{\tau'} |f_{1F}(\tau') - f_{2F}(\tau')| \leq q \sup\limits_{\tau} |f_1(\tau) - f_2(\tau)| \tag{5.54}$$

(where $f_{iF} = F(f_i)$, $i = 1, 2$) with a constant $q < 1$.

Proof. [Proof of Lemma] Let $f_i \in B(r_*)$ for $i = 1, 2$. The map F transforms the curves $x = f_i(\tau)$ onto the curves $x' = f_{iF}(\tau')$. Their parametric representations are

$$\tau_i' = \tau_i + \mu\, g_b\big(\lambda f_i(\tau_i) + (1 - \lambda)A\sin\tau_i\big), \tag{5.55}$$

$$x_i' = h_b\big(\lambda f_i(\tau_i) + (1 - \lambda)A\sin\tau_i\big), \qquad \tau_i \in R, \quad i = 1, 2. \tag{5.56}$$

If $\tau_1' = \tau_2' = \tau'$, then $x_1' - x_2' = f_{1F}(\tau') - f_{2F}(\tau')$. Hence

$$|f_{1F}(\tau') - f_{2F}(\tau')| \leq$$

$$\leq l_h\lambda\big(|f_1(\tau_1) - f_1(\tau_2)| + |f_1(\tau_2) - f_2(\tau_2)|\big) + l_h(1 - \lambda)A|\sin\tau_1 - \sin\tau_2| \leq$$

$$\leq l_h\big(\lambda r_* + (1 - \lambda)A\big)|\tau_1 - \tau_2| + \lambda l_h \|f_1 - f_2\|, \tag{5.57}$$

where $||f|| = \sup_\tau |f(\tau)|$.

Now we estimate the value of $|\tau_1 - \tau_2|$ for $\tau_1' = \tau_2'$, i.e. for

$$-(\tau_1 - \tau_2) = \mu g_b\big(\lambda f_1(\tau_1) + (1-\lambda)A\sin\tau_1\big) - \mu g_b\big(\lambda f_2(\tau_2) + (1-\lambda)A\sin\tau_2\big). \tag{5.58}$$

From the above equality it follows that

$$|\tau_1 - \tau_2| \le \mu l_g\big(\lambda r_* + (1-\lambda)A\big)|\tau_1 - \tau_2| + \mu\lambda l_g||f_1 - f_2||,$$

and hence

$$|\tau_1 - \tau_2| \le \frac{\mu\lambda l_g||f_1 - f_2||}{1 - \mu l_g\big(\lambda r_* + (1-\lambda)A\big)}. \tag{5.59}$$

Eventually, we conclude from (5.57) and (5.59) that

$$|f_{1F}(\tau') - f_{2F}(\tau')| \le q\sup_\tau |f_1(\tau) - f_2(\tau)|, \tag{5.60}$$

where

$$q = \frac{\lambda l_h}{1 - \mu l_g\big(\lambda r_* + (1-\lambda)A\big)} = \frac{\lambda R(r_*)}{\lambda r_* + (1-\lambda)A} < 1. \tag{5.61}$$

This finishes the proof of the lemma. □

Now we can continue the proof of Theorem 5.2.

The set $B(r_*)$ with the metric $\rho(f_1, f_2) = \sup_\tau |f_1(\tau) - f_2(\tau)|$ is a complete metric space and F is a contracting map in $B(r_*)$. By the Banach fixed point theorem (see Ref. [24]), in the space $B(r_*)$ there exists exactly one element $f = F(f)$. For every $f_0 \in B(r_*)$ the sequence $\{f_n\}$, defined by $f_{n+1} = F(f_n)$, converges to f. The curve $x = f(\tau)$ describes the invariant manifold, the existence of which has been proved.

The function T defined by (5.47) has the positive derivative

$$\frac{dT(\tau)}{d\tau} = 1 + \mu g_b'\big(\lambda f(\tau) + (1-\lambda)A\sin\tau\big)\cdot\big(\lambda f'(\tau) + (1-\lambda)A\cos\tau\big) \ge$$

$$\ge 1 - \mu l_g\big(\lambda r_* + (1-\lambda)A\big) = \frac{\lambda l_h}{q} > 0.$$

So, the dynamics on the invariant manifold is described by the increasing function T.

This completes the proof of Theorem 5.2. □

In Theorem 5.2 we use no other properties of the functions h_b and g_b except its Lipschitz continuity with constants l_h, l_g. These constants depend on the parameters b, A (see Fig.5.1) and they increase as A increases. For example, if $b = 0$, then $l_h = 1$ and $l_g = 2\pi(1 - A)^{-2}$.

Figure 5.6 shows two families of curves (for two values of b and five values of λ) in the μA-plane. In the domain lying below these curves the condition (5.48) is satisfied and there exists a globally stable invariant manifold (5.44).

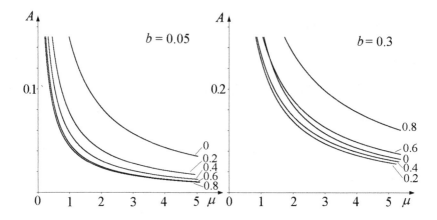

Fig. 5.6   Borders of a domain in which the invariant manifold exists for two values of b and for $\lambda = 0, 0.2, 0.4, 0.6, 0.8$. Below the given lines the condition (5.48) is satisfied.

5.3.2   Decay of the invariant manifold

If the map F given by (5.10)–(5.11) has the globally stable invariant manifold (5.44) then the dynamics is completely characterized by the rotation number ρ of the map (5.47). The map $T(\tau; \mu)$ defined by (5.47) is an increasing function of the parameter μ, and if it has T-*property* then the rotation number $\rho(\mu)$ is Cantor's step-function.

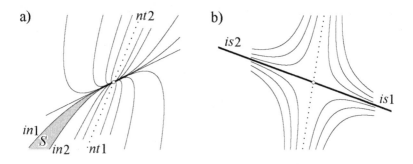

Fig. 5.7   The locally invariant curves of the map (5.34) in a neighborhood of fixed points: (a) of a stable node, and (b) of a saddle-type point.

If the number $\rho(\mu)$ is irrational and if $(\tau, x) \in M$, then the set of points $F^k(\tau, x)$ for $k = 1, 2, 3...$ is dense in the set M which is treated as a subset of cylinder $S \times R$, with the cyclic variable $\tau \bmod 2\pi$.

If $\rho(\mu) = \frac{n}{m}$ is a rational number then the map F has a periodic orbit of the type n/m on the manifold M. Usually, there is an even number of periodic orbits, and between any two stable nodes there exists a saddle--type periodic point, and vice versa. The invariant manifold is composed of periodic points and invariant segments of lines which connect saddle-type periodic points with stable nodes.

A family of locally invariant curves of the map (5.34) in a neighborhood of a stable node is shown in Fig.5.7a. All these curves, except exactly two $nt1$, $nt2$ (dotted lines in the figure), are tangent to the manifold M at this stable node. A family of locally invariant curves of the map (5.34) in a neighborhood of a saddle-type periodic point is shown in Fig.5.7b. Exactly two curves $is1$, $is2$ (bold line in the figure) lie on the manifold. If $(\tau, x) \in M$ then the sequence of points $G^k(\tau, x)$, where $G(\tau, x) = F^m(\tau, x) - (2\pi n, 0)$, tends to a node as $k \to \infty$ and tends to a saddle-type point as $k \to -\infty$.

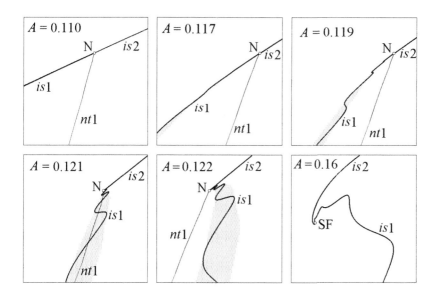

Fig. 5.8 Invariant segments $is1$ and $is2$ of the map (5.10)–(5.11) in a neighborhood of the node N on cylinder $S \times R$, for parameters $b = 0$, $\lambda = 0.1$, $\mu = 1.09$.

If we change parameters, for example if the value of A increases, then the invariant manifold M can be destroyed, but the invariant attractive set $M = \bigcap_{k=0}^{\infty} F^k(D_A)$ will remain. There exist several ways of decay of the invariant manifold (5.44). One of them is related to its behavior in the neighborhood of the node.

Let S denote a sector bounded by two locally invariant curves $in1$, $in2$ attached to the node (grey domain in Fig.5.7a). Always there exists such a sector S that the invariant curve $is1$ (or $is2$) starting in a saddle-type point (bold line in Fig.5.7b) crosses all locally invariant curves which belong to S (only exceptionally S reduces to one curve $in1 \equiv in2$).

An example is shown in Fig.5.8. For the map (5.10)–(5.11) with $b = 0$ (compare with (5.9)), $\lambda = 0.1$, $\mu = 1.09$, the invariant curves $is1$ and $is2$ are shown in a small neighborhood of the node N for several values of the parameter A. By Theorem 5.2, the smooth invariant manifold (5.44) exists for $A < 0.066$, but we observe that it probably exists for $A < 0.117$. For $A > 0.119$ the invariant attractive set cannot bee described by the equation $x = f(\tau)$ with the Lipschitzian function f; the manifold (5.44) does not exist. For greater values of A (e.g. for $A = 0.16$) the node N is replaced by a stable focus SF.

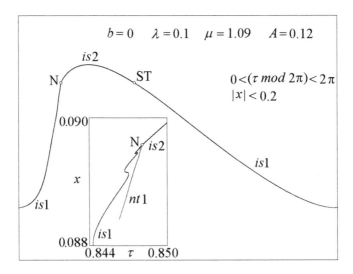

Fig. 5.9   An example of a globally stable invariant set of the map (5.10)–(5.11) and picture of a neighborhood of the node enlarged 100 times in vertical and 1000 times in horizontal directions.

An example of a globally stable invariant set composed of a node N, a saddle-type point ST and two invariant segments $is1$, $is2$ is shown in Fig.5.9. It looks like a graph of a smooth function $x = f(\tau)$, but an enlarged picture of a neighborhood of the node indicates that it is not a smooth manifold (5.44).

5.4 Strange attractors and chaotic steady-states

In this section selected properties of strange attractors and steady-state chaotic dynamics of the DPLL system will be presented.

Dynamics defined by the map (5.10)–(5.11) will be discussed. For simplicity, we assume $b = 0$. Then, the map $F : (\tau, x) \to (\tau', x')$ is given by equations

$$x' = \lambda x + (1 - \lambda)A \sin \tau, \qquad \tau' = \tau + \mu \frac{2\pi}{1 + x'}, \qquad (5.62)$$

where $0 < \lambda < 1$ and $0 < A < 1$.

5.4.1 *Maximal invariant set*

Let $C = \{(\tau, x) : x = c, |c| < A\}$, denote a horizontal line in the plane τ, x. Using geometrical properties of the map F, described in Section 5.1, we conclude that for every natural number k the curve $F^k(C)$ separates the plane into two simple connected sets. Examples of these curves for $k = 1, 2$ are shown in Fig.5.10 for three values of the parameter μ. The domains $F^k(D_c)$, where $D_c = \{(\tau, x) : x > c\}$, are marked in grey. If μA is sufficiently small, then the curves $F^k(C)$ tend to a smooth invariant manifold as $k \to \infty$, but if the values of μA are large, then the curves $F^k(C)$ are folded.

It is known that for each initial point (τ_0, x_0) there exists a number n such that $F^k(\tau_0, x_0)$ belongs to the set $D_A = \{(\tau, x) : |x| \le A\}$ for each $k > n$. From the relation $F(D_A) \subset D_A$ it follows that $F^{k+1}(D_A) \subset F^k(D_A)$ for $k = 1, 2, 3...$ The sequence of closed sets

$$D_A \supset F(D_A) \supset F^2(D_A) \supset \ldots \supset F^k(D_A) \supset \ldots \qquad (5.63)$$

tends to a nonempty closed set

$$D_\infty = \bigcap_{k=0}^{\infty} F^k(D_A), \qquad (5.64)$$

which is called the *maximal invariant set* (see [50]). Evidently, $F(D_\infty) = D_\infty$. The set (5.64) attracts all trajectories i.e. dist $[F^k(\tau_0, x_0), D_\infty] \to 0$ as $k \to \infty$ for each (τ_0, x_0).

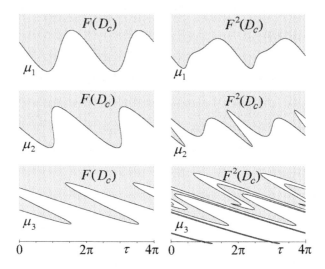

Fig. 5.10   Examples of domains $\{F^k(\tau, x) : x > 0\}$ for $k = 1, 2$ and for three values of the parameter μ (where $A = 0.9$, $\lambda = 0.95$, $\mu_1 = 1.6$, $\mu_2 = 4.5$, $\mu_3 = 17.2$).

On cylinder $S \times R$ the measure of the set $F^k(D_A)$ is equal to $4\pi A \lambda^k$ (because $J(\tau, x) = \lambda$ for $b = 0$) and, consequently, the measure of D_∞ is equal to zero.

For every natural k the set $F^k(D_A)$ is simply connected and it cuts the cylinder. The invariant set D_∞ has the same properties. The set $(S \times R) \backslash D_\infty$ has two disjoint open components. However, the geometrical structure of the set D_∞ can be very complicated. Two examples are shown in Fig.5.11. Maximal invariant set contains all bounded invariant sets of the map F, e.g. the periodic orbits, the invariant manifold, the invariant lines repelled from periodic saddle-type points, invariant sets of Smale's horseshoes and maybe other ones. If the map (5.62) has the globally stable invariant manifold (5.44), then the maximal invariant set coincides with this manifold.

5.4.2   Attractors

We are especially interested in attractors, i.e. the minimal invariant and attracting sets.

Definition 5.3. The closed invariant set $I = F(I)$ is called an attractor of the map F if it has a dense trajectory and if it is an attracting set, i.e.

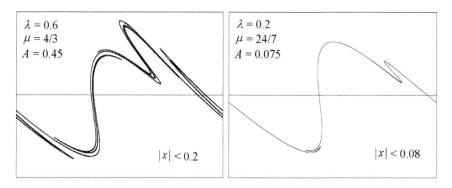

Fig. 5.11 Two examples of maximal invariant sets. On the horizontal axis $(\tau \bmod 2\pi) \in (0, 2\pi)$.

for every $\varepsilon > 0$ there exists $\eta > 0$ such that from $\mathrm{dist}\,[(\tau, x), I] < \eta$ it follows that $\mathrm{dist}\,[F^k(\tau, x), I] < \varepsilon$ for $k = 1, 2, \ldots$ and $\mathrm{dist}\,[F^k(\tau, x), I] \to 0$ as $k \to \infty$.

The existence of a dense trajectory in the set I means that the attractor has no proper invariant and attracting subset.

Sometimes an attractor is identical with the maximal invariant set (e.g. both graphs in Fig.5.11), but in general, attractors are proper subsets of the maximal invariant set. Let us distinguish three types of attractors:

– stable periodic orbits,
– globally stable Lipschitzian invariant manifold $x = f(\tau)$,
– strange attractors (all attractors which are not of the first two types).

If there exists a globally stable invariant manifold (5.44) then only two types of attractors are possible: this manifold (if the map (5.62) restricted to the manifold has an irrational rotation number) or stable periodic orbits. If the invariant manifold does not occur then stable periodic orbits and strange attractors are possible.

In Fig.5.12 a collection of selected strange attractors of the map (5.62) is shown for fixed values of parameters λ, μ and for several values of A. These attractors were calculated as a steady-state trajectory $F^k(\tau_0, x_0)$ (for $k = 50000, \ldots 100000$) of a random initial point (τ_0, x_0).

Geometrical properties of attractors depend on parameters — mostly in a very complicated form. An example shown in Fig.5.13 illustrates the "evolution" of an attractor for increasing value of A. For each A_i from

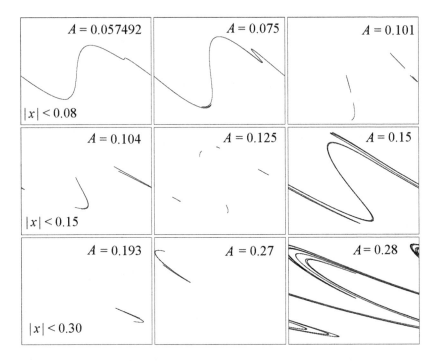

Fig. 5.12   Collection of attractors for $\lambda = 0.2$, $\mu = 3.428$ and for several values of the parameter A. On the horizontal axis $(\tau \bmod 2\pi) \in (0, 2\pi)$.

the set of 1200 uniformly distributed values of the parameter $A \in (0, 0.3)$ (i.e. for $A_i = 0.00025i$) the trajectory $F^k(\tau_0, x_0)$ is calculated for $k = 1, 2, ..., 1000$ and projected on both axes for $k = 500, ..., 1000$. We take $(\tau_{0,i+1}, x_{0,i+1}) = F^{1000}(\tau_{0,i}, x_{0,i})$, where $(\tau_{0,i}, x_{0,i})$ denotes the initial point of the trajectory calculated for A_i.

Two enlarged fragments of Fig.5.13a are shown in Fig.5.14. For a small value of the parameter A there exists a globally stable invariant manifold $M \subset S \times R$ and the map F restricted to this manifold is an increasing function. The steady-state trajectories are dense on the manifold M (e.g. if the rotation number is irrational) or there exist stable and unstable periodic orbits. A stable periodic orbit appears and disappears as a result of the saddle-node bifurcation (e.g. the orbit of the period 7 for $A \in (0.040, 0.046)$ in Fig.5.13).

If A increases, we observe decay of the invariant manifold similar to that described in Section 5.3.2. If we observe a period doubling bifurcation (e.g.

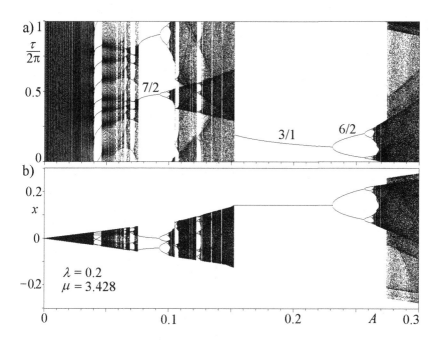

Fig. 5.13　Projections of attractors on the ($\frac{\tau}{2\pi}$ *mod* 1)-axis and on the x-axis for $A \in (0, 0.3)$.

for the value $A \approx 0.057491$ in Fig.5.14b) then the invariant manifold does not exist. Indeed, if the maximal invariant set D_∞ is a line then the map F restricted to this line is an increasing function because in the opposite case there would exist a point $(\tau, x) \in D_\infty \subset D_A$ which would have at least two preimages and this would contradict the fact that F has an inverse map F^{-1} in D_A. On the other hand, the one-dimensional map F increasing on the line D_∞ has only saddle-node bifurcations of periodic orbits. So, the period doubling bifurcation excludes the existence of an invariant manifold. In this case the attractor is not a line. However, it can be well approximated by a line or by several segments of a line (e.g. attractors shown in Fig.5.12 for $A = 0.057492$ and for $A = 0.125$).

It is easy to explain why the strange attractor can be well approximated by a line. Assume that λ is a fixed small value. For given values μ and A let the map F have an m-periodic orbit (stable or unstable). Each point (τ_*, x_*) of this orbit is a fixed point of the map $G = F^m$. Denote by ν_-, ν_+ eigenvalues of the matrix $G'(\tau_*, x_*)$. If m takes a large value then the determinant $\det G'(\tau_*, x_*) = \nu_- \cdot \nu_+ = \lambda^m$ is very small. For example, in

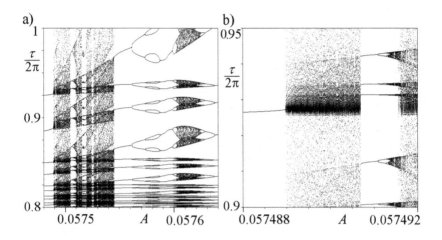

Fig. 5.14   Enlarged fragments of Fig.5.13a.

Fig.5.12 for $A = 0.125$, $\lambda = 0.2$, there is shown a strange attractor which contains an unstable orbit of period $m = 5$ and, consequently, $\lambda^m = 0.00032$. One of two eigenvalues, say ν_-, is very small in comparison with ν_+. It means that G maps a neighborhood of the point (τ_*, x_*) onto a set close to such a line L which is tangent at the point (τ_*, x_*) to the eigenvector related to ν_+. The area of this neighborhood is reduced λ^{-m}-times by the map G.

Changing the value of the parameter A we observe a sequence of period doubling bifurcations. Projections of attractors of the map F^m onto τ-axis (or x-axis) look as in Fig.4.39. For a value $A = A_{\text{sn}}$ there appears a stable fixed point (i.e. m-periodic orbit of F) by the saddle-node bifurcation. Next we observe sequences of period doubling bifurcations, the occurrence of continuous invariant measures and, eventually, the jump to another attractor for a value $A = A_Q$.

Several such examples of period m and interval (A_{sn}, A_Q) for $\lambda = 0.2$ and $\mu = 3.428$ are given below.

$1°$	$m = 30$	$A_{\text{sn}} = 0.0589$	$A_Q = 0.0592$
$2°$	$m = 23$	$A_{\text{sn}} = 0.0601$	$A_Q = 0.0604$
$3°$	$m = 16$	$A_{\text{sn}} = 0.0616$	$A_Q = 0.0621$
$4°$	$m = 30$	$A_{\text{sn}} = 0.0626$	$A_Q = 0.0622$
$5°$	$m = 9$	$A_{\text{sn}} = 0.0645$	$A_Q = 0.0661$
$*6°$	$m = 7$	$A_{\text{sn}} = 0.0691$	$A_Q = 0.0728$
$*7°$	$m = 2$	$A_{\text{sn}} = 0.0755$	$A_Q = 0.105$

$$
\begin{array}{llll}
*8° & m = 3 & A_{\mathrm{sn}} = 0.103 & A_Q = 0.108 \\
*9° & m = 5 & A_{\mathrm{sn}} = 0.122 & A_Q = 0.126 \\
10° & m = 3 & A_{\mathrm{sn}} = 0.1420 & A_Q = 0.1424 \\
*11° & m = 1 & A_{\mathrm{sn}} = 0.1425 & A_Q = 0.273
\end{array}
$$

The examples marked by asterisk can be seen in Fig.5.13. The others would be seen after stretching the scale of A-axis.

Results of a numerical experiment shown in Fig.5.13 give only one attractor for each A_i. In order to show other attractors we must change the algorithm of the choice of initial values $(\tau_{0,i}, x_{0,i})$.

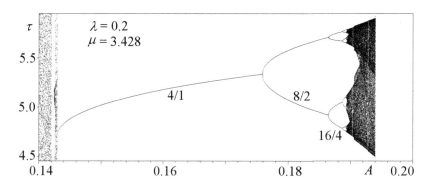

Fig. 5.15   Projections of attractors on the $(\tau \bmod 2\pi)$-axis versus A.

For certain intervals of the parameter A there exist more then one attractor. For example, if $A \in (0.103, 0.105)$ then there exists a stable orbit of the period 3 and a strange attractor containing two simply connected components. This attractor is shown in Fig.5.12 for $A = 0.104$. Similarly, for $A \in (0.1425, 0.1525)$ there exists a stable fixed point and a strange attractor shown in Fig.5.12 for $A = 0.15$.

Fig.5.15 shows the projection of an attractor on τ-axis for $\lambda = 0.2$, $\mu = 3.428$ and for $A \in (0.14, 0.20)$. This attractor coexists with a fixed point of the type $3/1$ and, moreover, with a strange attractor for $A \in (0.1425, 0.1525)$. For a selected value $A = 0.193$ the attractor is shown in Fig.5.12. The numbers $4/1$, $8/2$, $16/4$ in Fig.5.15 and the numbers $7/2$, $3/1$, $6/2$ in Fig.5.13 denote the types of periodic orbits. It is evident that a periodic orbit of the map F of the type n/m has m periodic points on cylinder $S \times R$ with cyclic variable $\tau \bmod 2\pi$.

5.4.3 *Attractive domains*

Attractive domains of periodic orbits and of strange attractors depend on the arrangement of invariant lines of saddle-type periodic points. Representative examples are shown in Fig.5.16 for $\lambda = 0.2$, $\mu = 24/7$ and for two values of the parameter A.

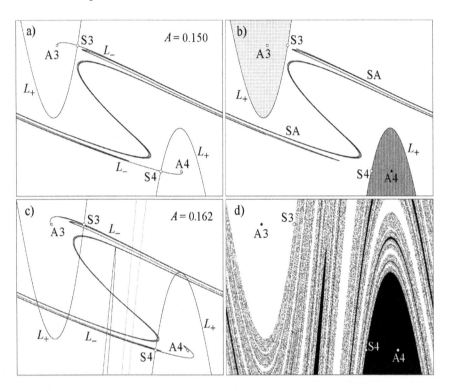

Fig. 5.16   Invariant lines of the saddle-type fixed points S3, S4 and attractive domains of the attractors A3, A4, SA. Values of parameters: $\lambda = 0.2$, $\mu = 24/7$, $(\tau \bmod 2\pi) \in$ $\in (0, 2\pi)$, $|x| < 0.2$.

Fig.5.16a (for $A = 0.15$) shows two saddle-type fixed points S3, S4 and two stable fixed points A3 and A4 (of types $3/1$ and $4/1$ respectively). Each saddle-type fixed point has two invariant lines L_- repelled from this point. One of them tends to the stable fixed point (to A3 or to A4 respectively) and the second tends to a strange attractor SA. Two invariant lines L_+ attracted to the saddle-type fixed point form the boundary of the attractive domain of the stable fixed point. For $A = 0.15$ there exist three attractors, two fixed points A3, A4 and the strange attractor SA.

In Fig.5.16b their attractive domains are marked in light grey, dark grey and white respectively. The steady-state stable output signal of the DPLL system is either a periodic of the period 3-times or 4-times greater then the period of the input signal, or a chaotic one.

If A increases then the distance from the strange attractor SA to the attractive domains of the points A3, A4 decreases, and homoclinic trajectories can occur. For a bifurcation value of the parameter A the homoclinic trajectory of the point S3 (or S4) appears, the strange attractor SA disappears and almost all points of its basin are now attracted to A3 (or to A4 respectively). This phenomenon is called the *crisis bifurcation* of the strange attractor. In the DPLL system the stable chaotic output signal is then replaced by a transient chaos. Fig.5.16c (for $A = 0.162$) shows the invariant lines L_- and small segments of lines L_+ for the case where both saddle-type fixed points S3 and S4 have homoclinic trajectories. Consequently, the attractive domains of the points A3 (white domain) and A4 (black domain) look as in Fig.5.16d. The border of these domains is formed by the invariant lines L_+ attracted to the saddle-type points S3 and S4.

Other examples are shown in Fig.5.17 for $\lambda = 0.4$, $\mu = 4/3$ and for two values of A. For $A = 0.411733$ there exist two stable fixed points A1, A2 and two saddle-type fixed points S1, S2 of the types 1/1 and 2/1 respectively. The invariant lines L_- and L_+ repelled from and attracted to the points S1, S2 are shown in Fig.5.17a. The parameter A takes the value very close to the bifurcation value A_{bh} of the homoclinic orbit of the point S1 (from numerical experiment it follows that $0.411733 < A_{bh} < 0.411734$).

If $A < A_{bh}$ then there exist three attractors shown in Fig.5.17b, i.e. two stable fixed points A1, A2 and the strange attractor SA. The steady-state stable output signal of the DPLL system is either a periodic of the input signal period or of the period 2-times greater, or a chaotic one. Attractive domains of these attractors are marked in grey, black and white respectively. The invariant line L_+ attracted to the point S2 from above and the invariant line L_- repelled from S2 to the right intersect along a homoclinic trajectory. This determines a very complicated border of the attractive domain of the point A2.

The structure of the strange attractor SA in a small neighborhood of the point S1 is shown in Fig.5.18 for $A < A_{bh}$. Directions of the invariant lines L_- (repelled from S1 to the right) and L_+ (attracted to S1 from below) are given by eigenvectors of the Jacobian matrix of the map (5.62) at the

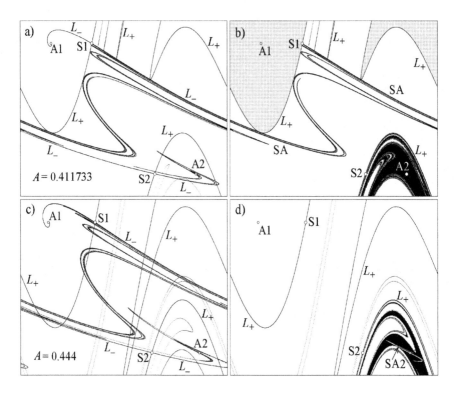

Fig. 5.17   Invariant lines of the saddle-type fixed points S1, S2 and attractive domains of the attractors A1, A2, SA. Values of parameters: $\lambda = 0.4$, $\mu = 4/3$, $(\tau \bmod 2\pi) \in$ $\in (0, 2\pi)$, $|x| < 0.45$.

point S1. According to (5.29), the eigenvalues ν_- and ν_+ of this matrix determine the behavior of trajectories $\{F^n(\tau, x)\}$ in a small neighborhood of the point S1. The points of a trajectory move along a locally invariant curve of a family shown in Fig.5.7b.

For $A > A_{\mathrm{bh}}$ there exists a homoclinic trajectory of the point S1. The invariant line L_+ attracted to S1 from below crosses the invariant line L_- repelled from S1 to the right (Fig.5.17c). The strange attractor SA disappeared and almost all points, which were attracted to it, are now attracted to the fixed point A1 (Fig.5.17d).

For $A = 0.444$ the fixed point A2 is of the saddle-type and it belongs to the strange attractor SA2 with the attractive domain marked in black in Fig.5.17d. The enlarged detail of Fig.5.17d which contains the strange attractor SA2 is shown in Fig.5.19 in inverse colors. If an initial value

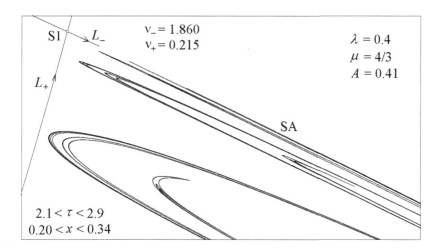

Fig. 5.18 Structure of the strange attractor SA in a neighborhood of the saddle-type fixed point S1 of the type 1/1.

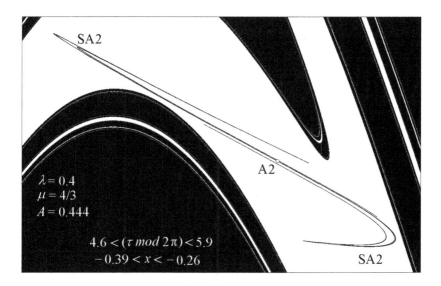

Fig. 5.19 Strange attractor SA2 and its attractive domain (white domain). This is an enlarged detail of Fig.5.17d in inverse colors. Black domain belongs to attractive set of the fixed point A1.

belongs to the attractive domain of SA2 (white domain in Fig.5.19) then the steady-state stable output signal of the DPLL system is chaotic, but its period mean value is exactly 2-times greater then the period of the input signal.

The "evolution" of the attractor SA2 is similar to that shown in Fig.5.15. For $A = A_{sn}$ the pair of fixed points S2 and A2 appears as a result of the saddle-node bifurcation. If A increases then the distance from S2 to A2 increases, and the attractive domain of the stable point A2 grows in size. For a value $A = A_{pdb}$ the point A2 loses its stability (it becomes a saddle-type fixed point) by the period doubling bifurcation. The attractor changes from a stable point into a stable orbit of period two. Next we observe the Feigenbaum cascade of the period doubling bifurcations and the occurrence of a strange attractor SA2. This attractor disappears (crisis bifurcation) for $A = A_{bh}$ when it touches the invariant line L_+ attracted to S2 (the set SA2 touches the black domain in Fig.5.19). For $A > A_{bh}$ almost all points, which were attracted to SA2, are now attracted to A1.

It can be shown that if there exist two stable fixed points of the types $n/1$ and $(n + 1)/1$ than always a transient chaos or a steady-state stable chaos (strange attractor) is observed in a domain of initial values.

Bibliography

[1] Adler R.: *A study of locking phenomena in oscillators*. Proc. IRE, Vol. 34, June 1946, pp. 351-357.

[2] Andronov A.A., Khaikin S.E., Witt A.A.: *Theory of Oscillations*. Pergamon Press, New York, 1966.

[3] Anosov D.V., Arnold V.I. (eds.): *Dynamical systems I* Encyclopedia of Mathematical Sciences, Vol.1 Springer-Verlag, Berlin, Heidelberg 1988.

[4] Arnold V.I.: *Geometric methods in the theory of ordinary differential equations*. Springer-Verlag, New York, Heidelberg, Berlin 1983.

[5] Arnold V.I.: Remarks on perturbation theory for problems of the Mathieu type. (Russian) Usp. Mat. Nauk 38, No.4 (232), 189-203 (1983). English transl.: Russ. Math. Surv. 38, No.4, 215-233 (1983). Zbl.541.34035.

[6] Arrowsmith D.K., Place C.M.: *An introduction to dynamical systems*. Cambridge University Press, Cambridge, New York 1990.

[7] Bamon R., Malta I.P., Pacifico M.J., Takens F.: Rotation intervals of endomorphisms of the circle. Erg. Th. Dyn. Sys., 4, 1984, pp. 493-498.

[8] de Bellescize H.: La reception synchrone. Onde electr., Vol. 11, June 1932, pp. 230-240.

[9] Best R.E.: *Phase-locked loops*. McGraw-Hill, New York 1984.

[10] Blanchard A.: *Phase-locked loops*. J. Wiley, New York, London, Sydney, Toronto, 1976.

[11] Blanchard P.: Complex analytic dynamics on the Riemann sphere. Bull. Am. Math. Soc. Vol. 11, n. 1, July 1984, pp. 85-141.

[12] Block L., Guckenheimer J., Misiurewicz M., Young L.S.: Periodic points and topological entropy of one dimensional maps. Lect. Notes Math. 819, Springer-Verlag, 1980, pp. 18-34.

[13] Bogoliubov N.N., Mitropolskii Y.A.: *Asymptotic methods in the theory of nonlinear oscillations*. Gordon and Breach, New York 1961.

[14] Butenin N.V., Neimark Ju.A., Fufaev N.A.: *Introduction to the theory of nonlinear oscillations*. (Russian) Nauka, Moscow 1976.

[15] Collet P., Eckmann J.P.: *Iterated maps on the interval as dynamical systems.* Birkhäuser, Boston, Basel, Stuttgart 1980.

[16] Devaney R.L.: *An introduction to chaotic dynamical systems.* Addison-Wesley Publishing Company, Inc., 1989.

[17] Endo T., Chua L.O., Narita T.: Chaos from phase-locked loop – Part II Non-Hamiltonian case. IEEE Trans. Circ. Syst., Vol. CAS 36, February 1989.

[18] Feigenbaum M.J.: Quantitative universality for a class of nonlinear transformations. J. Stat. Phys. 19 (1978), pp. 25-52, 21 (1979), pp. 669-706.

[19] Gardner F.M.: *Phaselock Techniques.* J. Wiley, New York, Chichester, Brisbane, Toronto, 1979.

[20] Guckenheimer J., Holmes P.: *Nonlinear oscillations, dynamical systems and bifurcations of vector fields.* Springer-Verlag, 1983.

[21] Guckenheimer J., Moser J., Newhouse S.E.: *Dynamical systems.* CIME Lectures, Bressanone, Italy, June 1978, Birkhäuser, Boston 1980.

[22] Gupta S.C.: *Phase-locked loops.* Proc. IEEE, Vol. 63, February 1975, pp. 291-306.

[23] Hale J.K.: *Oscillations in nonlinear systems.* McGraw-Hill, New York 1963.

[24] Hale J.K.: *Ordinary differential equations.* Robert E. Kreiger Publ. Comp. INC., New York 1980.

[25] Hayashi Ch.: *Nonlinear oscillations in physical systems.* McGraw-Hill, Inc., 1964.

[26] Jakobson M.V.: On the number of periodic trajectories for analytic difeomorphisms of circles. Funct. anal. appl. (Russian), 1985, vol.19 No.1.

[27] Katok A., Hasselblatt B.: *Introduction to the modern theory of dynamical systems.* Encyclopedia of mathematics and its applications. Vol. 54, Cambridge University Press, 1995.

[28] Klapper J., Frankle J.T.: *Phase-locked and frequency-feedback systems.* Academic Press, New York, London 1972.

[29] Kudrewicz J.: *Dynamika pętli fazowej (Dynamics of Phase-Locked Loops).* Wydawnictwa Naukowo Techniczne, Warszawa 1991.

[30] Kudrewicz J., Grudniewicz J., Świdzińska B.: Chaotic oscillations as a consequence of the phase-slipping phenomenon in a discrete phase-locked loop. Int. Symp. on Circuits and Systems , San Jose, California, May 1986.

[31] Kudrewicz J.: *Nieliniowe obwody elektryczne (Nonlinear electrical circuits).* Wydawnictwa Naukowo Techniczne, Warszawa 1996.

[32] Li T.Y., Yorke J.A.: Period three implies chaos. Amer. Math. Mon., 82, 1975, pp.985-992.

[33] Lindsey W.C.: *Synchronization systems in communication and control.* Prentice-Hall, Englewood Cliffs 1972.

[34] Lindsey W.C., Chie C.M.: A survey of digital phase-locked loops. Proc. IEEE, Vol. 69, No. 4, April 1981, pp.410-431.

[35] Lindsey W.C., Simon M.K. (eds.): *Phase-Locked Loops and Their Application.* IEEE Press 1978.

[36] MacKay R.S., Tresser C.: Transition to topological chaos for circle maps. Physica, 19D, 1986, pp. 206-237.

[37] MacKay R.S., Tresser C.: Some flesh on the skeleton. The bifurcation structure of bimodal maps. Physica, 27D, 1987, pp. 412-422.

[38] Mané R.: *Ergodic theory and differentiable dynamics.* Springer-Verlag, Berlin, Heidelberg 1987.

[39] May R.M.: Simple mathematical models with very complicated dynamics. Nature, 261, 1976, pp. 459-467.

[40] De Melo V., van Strien S.: *One-dimensional dynamics.* Springer-Verlag, Berlin, Heidelberg 1993.

[41] Misiurewicz M., Szlenk W.: Entropy of piece-wise monotone mappings. Studia Math. 67, 1980.

[42] Mitropolski A. Yu.: *Averaging method in nonlinear mechanics.* (Russian) Naukova Dumka, Kiev 1971.

[43] Moser J.: *Stable and random motions in dynamical systems.* Princeton Univ. Press, Princeton 1973.

[44] Nitecki Z.: *Differentiable dynamics.* MIT Press Cambridge, Massachusetts, London 1971.

[45] Odyniec M., Chua L.O.: Josephson-junction circuit analysis via integral manifolds. IEEE Transactions, CAS 30 May 1983, pp.308-320, Part II CAS 32 January 1985, pp. 34-45.

[46] Osborne H.C.: Stability analysis of n-th power digital phase-locked loop. Part I, II, IEEE Trans. Commun., Vol. COM 28, August 1980, pp. 1343-1364.

[47] Ostlund S., Rand D., Siggia E.: Universal properties of the transition from quasi-periodicity to chaos in dissipative systems. Physica, 8 D, 1983, pp. 303-342.

[48] Palis J.Jr., de Melo V.: *Geometric theory of dynamical systems. An introduction.* Springer-Verlag, New York, Heidelberg, Berlin 1982.

[49] Palis J., Takens F.: Hyperbolicity and sensitive chaotic dynamics at homoclinic bifurcations. Cambridge studies in advanced mathematics 35, Cambridge University Press, 1993.

[50] Robinson C.: *Dynamical systems: stability, symbolic dynamics and chaos.* CRC Press, Inc. Boca Raton, Ann Arbor, London, Tokyo 1995.

[51] Ruelle D.: *Elements of differentiable dynamics and bifurcation theory.* Academic Press, Inc., London 1989.

[52] Ruelle D.: *Chaotic evolution and strange attractors.* Cambridge University Press, 1989.

[53] Shahgildian V.V., Belustina L.N. (eds.): *Systems of phase synchronization.* (Russian) Radio Svaz', Moscow 1982.

[54] Sharkovskii A.N.: Coexistence of the cycles of a continuous mapping of the line into itself. (Russian) Ukrain. Math. Zh. 16 (1), 1964, pp. 61-71.

[55] Sharkovskii A.N., Maistrenko Yu.L., Romanenko E.Ju.: *Difference equations and their applications.* (Russian) Naukova Dumka, Kiev 1986.

[56] Smale S.: Differentiable dynamical systems. Bull. Am. Math. Soc., 73, 1967, pp. 747-817.

[57] Szlenk W.: *An introduction to the theory of smooth dynamical systems.* John Wiley and Sons, Chichester 1984.

[58] Van Trees H.L.: *Detection, estimation and modulation theory.* John Wiley and Sons, Inc., New York, London, Sydney, Toronto 1982.

[59] Viterbi A.J.: *Principles of coherent communication.* McGraw-Hill, New York 1966.

[60] Walters P.: An introduction to ergodic theory. Graduate Texts in Mathematics 79, Springer-Verlag New York Inc., New York, Heidelberg, Berlin 1982.

[61] Wąsowicz S.: The model of an analog-to-digital converter with random switching element. Proceedings of the Polish-Czech-Hungarian Workshop on Circuit Theory, Signal Processing and Applications. Sept. 3–7, 1997 Budapest.

Index